# Solid Phase Extraction

# Solid Phase Extraction: State of the Art and Future Perspectives

Special Issue Editor

**Victoria Samanidou**

MDPI • Basel • Beijing • Wuhan • Barcelona • Belgrade

**MDPI**

*Special Issue Editor*
Victoria Samanidou
Aristotle University of Thessaloniki
Greece

*Editorial Office*
MDPI
St. Alban-Anlage 66
4052 Basel, Switzerland

This is a reprint of articles from the Special Issue published online in the open access journal *Molecules* (ISSN 1420-3049) from 2018 to 2019 (available at: https://www.mdpi.com/journal/molecules/special_issues/SPE)

For citation purposes, cite each article independently as indicated on the article page online and as indicated below:

LastName, A.A.; LastName, B.B.; LastName, C.C. Article Title. *Journal Name* **Year**, *Article Number*, *Page Range*.

**ISBN 978-3-03921-158-6 (Pbk)**
**ISBN 978-3-03921-159-3 (PDF)**

Cover image courtesy of Victoria Samanidou.

# Contents

# About the Special Issue Editor

**Victoria Samanidou** is a Professor of Analytical Chemistry at the School of Chemistry of the Aristotle University of Thessaloniki, Greece. She obtained her doctorate in Chemistry in 1990 from the same department. She has published more than 165 original research articles in peer-reviewed journals and 60 reviews and chapters in scientific books. Her H-index is 31 (Scopus, June 2019). She is a member of the editorial board of more than 10 scientific journals and has reviewed more than 500 manuscripts in around 100 scientific journals. She was a guest editor of several Special Issues in scientific journals. In 2016 she was recognized on the Power List 2016, Analytical Scientist Magazine, Texere Publishing, as one of the top 50 women with the high impact on Analytical Chemistry. In January 2016, she was elected the President of the Division of Central and Western Macedonia of the Greek Chemists' Association.

# Preface to "Solid Phase Extraction: State of the Art and Future Perspectives"

Sample preparation is without a doubt the most important step preceding sample analysis. It is the step that determines the accuracy and speed of the obtained result. Due to the fact that it usually involves more than one step, it is tedious, time consuming, and any potential errors will accumulate and yield an erroneous outcome. Thus, it is considered by all analytical chemists as the bottleneck of every method. Though no sample preparation would, ideally, be the best approach, an effective extraction step is often required, not to say inevitable. To this end, solid-phase extraction (SPE) has been a determinative player in the challenge of chemical analysis during the last decades. Meanwhile, many replacement sample preparation approaches promising better and "greener" performance have evolved. However, SPE still plays a key role in method development, and advanced sorbent technology can re-orient the traditional approach to new perspectives. This book is a collection of papers describing the state of the art nature of this sample preparation technique and presenting recent and future advances. Thirteen outstanding manuscripts are included in this Special Issue and from this point I wish to thank all authors for their fine contributions.

**Victoria Samanidou**
*Special Issue Editor*

*molecules*

MDPI

*Article*

# Graphene-Derivatized Silica Composite as Solid-Phase Extraction Sorbent Combined with GC–MS/MS for the Determination of Polycyclic Musks in Aqueous Samples

Cheng Li [1,2], Jiayi Chen [1,2], Yan Chen [1,2], Jihua Wang [1], Hua Ping [1] and Anxiang Lu [1,2,*]

[1]   Beijing Research Center for Agricultural Standards and Testing, Beijing Academy of Agriculture and Forestry Sciences, Beijing 100097, China; lic@brcast.org.cn (C.L.); chenjy@brcast.org.cn (J.C.); cheny@brcast.org.cn (Y.C.); wangjihua@brcast.org.cn (J.W.); pingh@nercita.org.cn (H.P.)
[2]   Beijing Municipal Key Laboratory of Agriculture Environment Monitoring, Beijing 100097, China
*   Correspondence: anxiang_lu@hotmail.com; Tel.: +86-10-51503057

Received: 6 January 2018; Accepted: 1 February 2018; Published: 2 February 2018

**Abstract:** Polycyclic musks (PCMs) have recently received growing attention as emerging contaminants because of their bioaccumulation and potential ecotoxicological effects. Herein, an effective method for the determination of five PCMs in aqueous samples is presented. Reduced graphene oxide-derivatized silica (rGO@silica) particles were prepared from graphene oxide and aminosilica microparticles and characterized by scanning electron microscopy, Fourier transform infrared spectroscopy, and X-ray photoelectron spectroscopy. PCMs were preconcentrated using rGO@silica as the solid-phase extraction sorbent and quantified by gas chromatography–tandem mass spectrometry. Several experimental parameters, such as eluent, elution volume, sorbent amount, pH, and sample volume were optimized. The correlation coefficient ($R$) ranged from 0.9958 to 0.9992, while the limits of detection and quantitation for the five PCMs were 0.3–0.8 ng/L and 1.1–2.1 ng/L, respectively. Satisfactory recoveries were obtained for tap water (86.6–105.9%) and river water samples (82.9–107.1%), with relative standard deviations <10% under optimal conditions. The developed method was applied to analyze PCMs in tap and river water samples from Beijing, China. Galaxolide (HHCB) and tonalide (AHTN) were the main PCM components detected in one river water sample at concentrations of 18.7 for HHCB, and 11.7 ng/L for AHTN.

**Keywords:** graphene; solid-phase extraction; polycyclic musks; water; GC–MS/MS

## 1. Introduction

Personal care products (PCPs) are an important class of emerging pollutants that have raised significant concerns because of their bioaccumulation ability and potential adverse effects on the ecological environment [1]. Polycyclic musks (PCMs) are a representative group of PCP compounds. PCMs are commonly used as fragrances in various consumer products, such as shampoos, body washes, and detergents [2]. PCMs can enter the water supply in effluents from municipal wastewater treatment plants. Because of their extensive use and increasing consumption worldwide in recent years, PCMs are ubiquitously detected in water environments [3–6] and even aquatic organisms [7]. PCMs have been found to bioaccumulate, with some studies suggesting that they could have ecotoxicological effects on specific organisms and cause endocrine disruption in humans [8–11]. A frequently used PCM, 1,3,4,6,7,8-hexahydro-4,6,6,7,8,8-hexamethylcyclopenta[γ]-2-benzopyran (HHCB), is among the top 50 high priority pollutants suggested by Howard and Muir, with persistence and bioaccumulation potential that require further monitoring [3,12]. As PCMs are not included in routine monitoring programs, data on their environmental impact remains insufficient [13]. To prevent adverse effects on the ecosystem, the development of a fast and sensitive method to determine these emerging organic pollutants in water is important.

Several instrumental methods have been developed for PCM determination, including high-performance liquid chromatography (HPLC) [14] and gas chromatography coupled with electron capture detection (GC–ECD), mass spectrometry (GC–MS) [15], or triple quadrupole tandem mass spectrometry (GC-MS/MS) [16,17]. GC–MS/MS can provide enhanced selectivity and sensitivity compared with conventional single quadrupole GC–MS, effectively eliminating matrix interferences. Owing to the trace PCM levels in water environments and the complexity of various water matrices, an efficient sample pretreatment step is crucial to eliminate matrix interference and concentrate the target analytes before analysis. Solid-phase extraction (SPE) is a superior and frequently used extraction technique for organic compounds in water samples. The sorbent material is a vital factor in determining the concentrating ability and recovery capacity of SPE [18]. To enrich hydrophobic organic pollutants (HOCs), reversed-phase SPE on hydrophobic or mixed-mode sorbents, such as C18-silica and *N*-vinylpyrrolidone polymer (Oasis HLB), have been commonly used [19].

Graphene is a novel two-dimensional carbonaceous material with superior chemical stability, excellent thermal stability, and an ultra-high specific surface area with a theoretical value of about 2630 m$^2$/g, suggesting a high sorption capacity and great potential for use as a sorbent material [20]. Because of this, graphene and its complexes have become attractive sorbents in sample preparation procedures for a variety of analytes, such as phthalate esters [21], pesticides [22], polycyclic aromatic hydrocarbons [23]. Several pretreatment methods based on graphene have also been developed, including SPE, solid-phase microextraction (SPME), and dispersive solid-phase extraction (dSPE) [22–29]. Recently, graphene complexes were also used for the determination of synthetic musks by SPME or dSPE methods [30,31]. Despite the fact that SPE is the most frequently used extraction technique, the reported graphene materials used for PCMs by the SPE method are still limited. As an SPE sorbent, single-use graphene may lead to irreversible graphene aggregation, which could then escape the SPE cartridge under high pressure and reduce the extraction efficiency [26,32]. The immobilization of graphene on a solid support, such as silica microparticles, polymer, or steel, is an effective method to solve this problem [19]. Some studies have shown that using graphene-supported silica as a sorbent obtained excellent extraction performance for the enrichment of various analytes.

This work aimed to develop an effective and sensitive method for PCM determination in aqueous samples. The hybrid material, reduced graphene oxide-derivatized silica (rGO@silica), was synthesized and used as a new SPE sorbent for the simultaneous preconcentration of five PCMs in water. After SPE, the target analytes were quantified by GC–MS/MS. Several parameters, such as eluting solvent, sorbent amount, pH, and sample volume, were investigated. Under the optimal conditions, this novel method was successfully used for PCM determination in water samples.

## 2. Results and Discussion

### 2.1. Characterization of rGO@silica

The morphology of as-prepared rGO@silica was characterized by SEM. Figure 1A,B show images of bare aminosilica microspheres. These aminosilica microparticles had a spherical shape and smooth surface. Figure 1C,D show SEM images of rGO@silica, which clearly shows that the aminosilica microspheres were tightly encapsulated by the reduced graphene oxide flakes with the typical semitransparent and crumpled sheet structure. This result suggested that the rGO layer was successfully immobilized on the surface of the aminosilica microspheres.

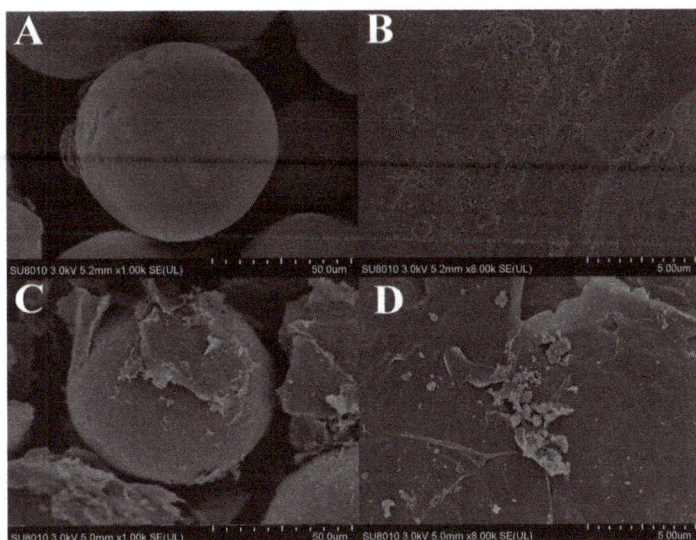

**Figure 1.** (**A**) SEM image and (**B**) high-magnification SEM image of aminosilica; (**C**) SEM image and (**D**) high-magnification SEM image of rGO@silica sorbent.

FT-IR spectra of aminosilica and rGO@silica are shown in Figure S1 (see Supplementary Materials). The main peaks at 798, 955, 1096, and 1640 $cm^{-1}$ were attributed to the SiO–H bending vibration, Si–OH stretching vibration, Si–O–Si stretching vibration, and N–H bending vibration, respectively [26]. The peaks at 3450 $cm^{-1}$ were assigned to the O–H stretching vibration. These results suggested that rGO was immobilized on the silica microsphere surface. The peak at 3450 $cm^{-1}$ for rGO@silica was stronger than that of aminosilica because the amount of OH groups in rGO@silica was larger. These characteristic spectra were consistent with those reported by Liu et al. [26]. Furthermore, rGO@silica was characterized by X-ray photoelectron spectroscopy (XPS) (Figure S2), which showed that the carbon peak intensity of rGO@silica was much higher than that of aminosilica. This evidence further confirmed the successful rGO immobilization on the silica surface.

## 2.2. Optimization of SPE Procedures

The sorption of PCMs on rGO@silica-packed SPE cartridge was investigated. When an aqueous solution (0.5 μg/mL) of five PCMs was passed through the cartridge, no target analytes were found in the outflow, suggesting a good retention ability for PCMs. To obtain the appropriate extraction efficiency, several experimental parameters, including elution solvent, sorbent amount, sample volume, and pH, were evaluated. The assessment was undertaken by loading rGO@silica cartridges with a standard mixture of five PCMs in water (200 mL, 100 ng/L containing 0.5% methanol). Optimization experiments were performed in duplicate and the mean values of the results were used.

### 2.2.1. Effect of the Elution Solvent

The choice of eluent is an important parameter determining the final extraction efficiency. The performances of *n*-hexane, dichloromethane (DCM), acetone, acetonitrile, and ethyl acetate as elution solvents were assessed. In each case, PCMs (20 ng of each) in aqueous solution (200 mL containing 0.5% methanol) were loaded onto the SPE cartridges, and different eluents were then passed through the cartridge. The elution efficiency of each solvent is shown in Figure 2A. DCM showed the best desorption capability for all five PCMs under the same extraction and elution conditions. Next, the volume of DCM was optimized by changing it from 4 to 12 mL over a series of tests. As shown in Figure 2B, the recoveries of the five PCMs increased with increasing DCM volume in the range 4–10 mL. Satisfactory recoveries were obtained with an eluent volume of 10 mL. Therefore, to achieve the best extraction performance, 10 mL of DCM was used in the subsequent experiments.

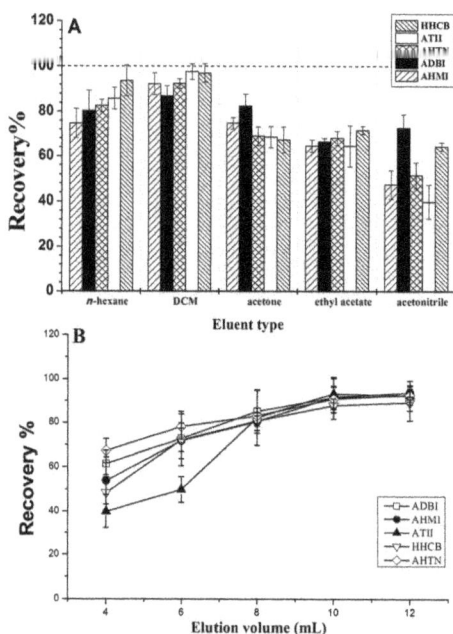

**Figure 2.** Optimization of (**A**) elution solvent and (**B**) elution volume for solid-phase extraction (SPE) of five polycyclic musks (PCMs) from water samples.

### 2.2.2. Effect of the Sorbent Amount

To ensure a sufficient analyte extraction efficiency, different amounts of rGO@silica (20–300 mg) were investigated. The same amount of PCMs (20 ng of each PCM in 200 mL of aqueous solution) was loaded to study the effect of the sorbent amount on the analyte recovery. As shown in Figure 3, as the amount of rGO@silica increased from 20 to 100 mg, the recoveries also increased. However, when the amount reached 300 mg, the recoveries slightly decreased. Therefore, 100 mg of rGO@silica was considered appropriate for the enrichment of these PCMs in water samples.

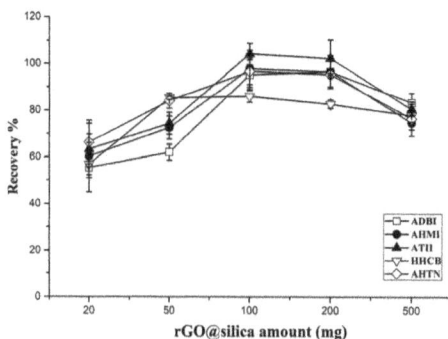

**Figure 3.** Effect of rGO@silica amount on SPE efficiency.

4

### 2.2.3. Effect of the Sample Volume and pH

Different volumes of aqueous solutions (50, 100, 200, 300, and 500 mL) spiked with 20 ng of each analyte were then investigated to determine the breakthrough volume, with recoveries shown in Figure 4A. Satisfactory recoveries (>85%) were obtained when the sample volume was below 300 mL. When the sample volume was increased to 500 mL, a partial sample loss or breakthrough occurred. For volumes below 500 mL, good recoveries were obtained. As tandem mass spectrometry provides enhanced sensitivity compared to conventional single quadrupole GC–MS, 200 mL of aqueous sample was able to achieve sufficient sensitivity for PCM analysis. Furthermore, a lower sample volume can reduce the extraction time. Therefore, a loading sample volume of 200 mL was used in the subsequent experiments.

The effect of the extraction pH on the recoveries was determined by adjusting the samples to various pH values ranging from 3 to 9 using 0.1 M HCl or 0.1 M NaOH. As shown in Figure 4B, no obvious variations were observed, and all five PCMs achieved good recoveries, ranging from 85.5% to 107.4%. This might be due to all PCMs being neutral under experimental conditions and their formation not being affected by different pH values. Therefore, the pH did not need to be adjusted in this study.

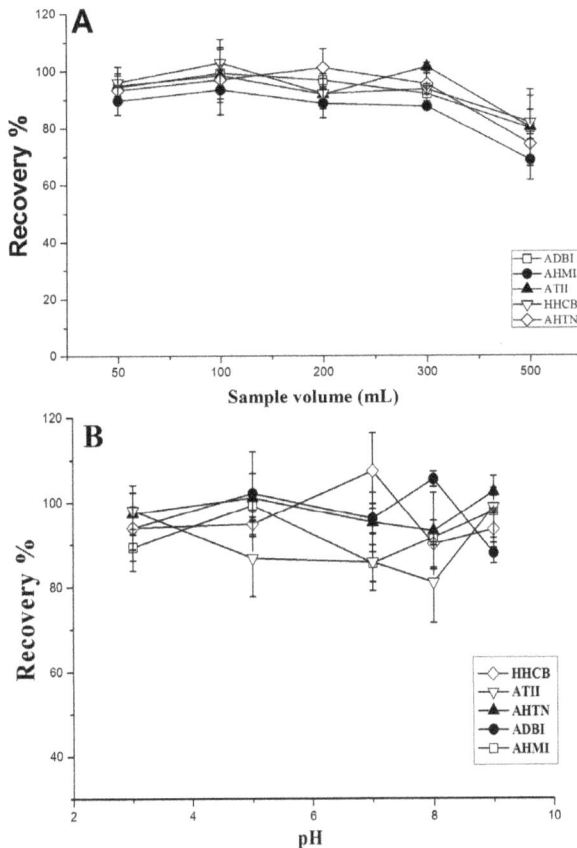

**Figure 4.** Effect of (**A**) sample volume and (**B**) solution pH on SPE efficiency.

### 2.3. Comparison with Other Sorbents

Because of their hydrophobic character, PCMs are commonly enriched using reverse-phase sorbents. To evaluate the enrichment capacity of rGO@silica, its performance was compared with those of conventional reserved-phase sorbent materials, including C18 silica and Oasis HLB, using the same amount (100 mg) of sorbent packed in 3 mL SPE cartridges. The cartridges were loaded with sample solutions (200 mL) spiked with the five PCMs (20 ng of each). The results are shown in Figure 5. rGO@silica yielded higher recoveries (>86%) than HLB, while C18 silica yielded the poorest recoveries (29.3–48.1%), suggesting that the sorption capacity of C18 silica was much weaker than that of rGO@silica. Therefore, rGO@silica is an excellent sorbent for PCM determination in water.

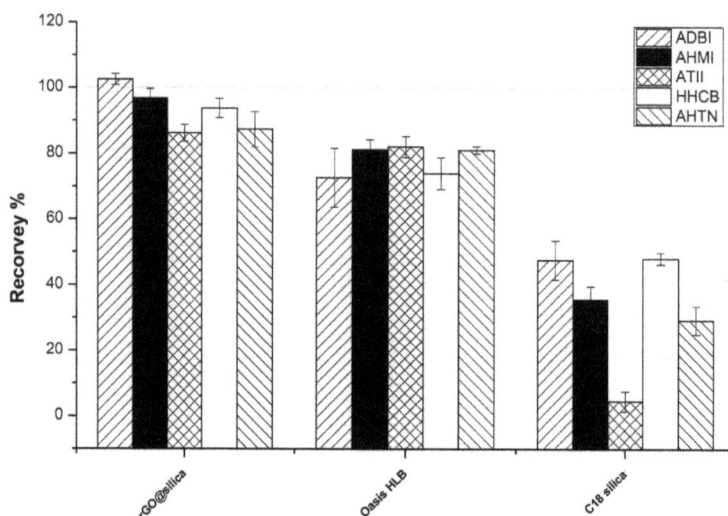

**Figure 5.** Comparison of the performance of rGO@silica with that of several other sorbents.

### 2.4. Method Validation

The linear regression, precision, limits of detection (LOD) and quantification (LOQ), repeatability, and recoveries of the developed method were investigated. The results are displayed in Table 1. The LOD and LOQ ranges for the five PCMs were 0.3–0.8 ng/L and 1.1–2.2 ng/L at signal-to-noise ratios of 3 and 10, respectively. The calibration curves were performed using six different concentrations (10, 20, 50, 100, 200, and 500 ng/L). A good linearity was obtained for the PCMs throughout the concentration range ($R > 0.99$).

**Table 1.** Analytical parameters for PCM determination using GC–MS/MS.

| Analyte | Linear Range (ng/L) | R | LOD (ng/L) | LOQ (ng/L) |
|---------|---------------------|--------|------------|------------|
| ADBI | 10–500 | 0.9992 | 0.5 | 1.5 |
| AHMI | 10–500 | 0.9978 | 0.3 | 1.1 |
| ATII | 10–500 | 0.9958 | 0.8 | 2.1 |
| HHCB | 10–500 | 0.9976 | 0.6 | 1.4 |
| AHTN | 10–500 | 0.9977 | 0.5 | 1.2 |

Two kinds of water samples (tap water and river water) were considered for the application of the proposed method. Repeatability and recovery studies were conducted for both tap and river water samples. Blank water samples containing no detectable PCMs were filtered using 0.75-μm

Whatman filter paper, while 200 mL samples of tap and river water were spiked with PCMs at 20, 100, and 200 ng/L in three parallel experiments. As shown in Table 2, the analyte recoveries in the fortified samples were 86.6–105.9% for tap water and 82.9–107.1% for river water. The relative standard deviations (RSDs) for all tap and river samples were below 10%. Figure S3 shows a typical total ion chromatogram (TIC) and multiple reaction monitoring (MRM) chromatogram of the tap water sample spiked with PCMs at 100 ng/L.

**Table 2.** Recovery studies on tap water and river water samples containing PCMs.

| Analyte | Spiked Levels (ng/L) | Tap Water Sample | | River Water Sample | |
|---|---|---|---|---|---|
| | | Recovery (%) | RSD (%) | Recovery (%) | RSD (%) |
| ADBI | 50 | 91.3 | 5.2 | 88.6 | 5.7 |
| | 100 | 102.4 | 6.1 | 102.3 | 3.5 |
| | 200 | 97.8 | 4.5 | 82.9 | 3.9 |
| AHMI | 50 | 89.4 | 1.9 | 87.1 | 2.6 |
| | 100 | 99.2 | 2.3 | 97.1 | 3.8 |
| | 200 | 92.3 | 0.8 | 93.1 | 5.2 |
| ATII | 50 | 99.1 | 2.1 | 96.9 | 5.8 |
| | 100 | 98.8 | 2.4 | 101.1 | 3.3 |
| | 200 | 89.6 | 3.4 | 85.3 | 5.9 |
| HHCB | 50 | 96.9 | 2.7 | 107.1 | 2.5 |
| | 100 | 93.1 | 5.2 | 106.3 | 3.3 |
| | 200 | 86.6 | 1.7 | 103.9 | 3.1 |
| AHTN | 50 | 93.3 | 5.9 | 99.7 | 5.7 |
| | 100 | 95.6 | 2.9 | 96.4 | 4.3 |
| | 200 | 105.9 | 5.5 | 84.5 | 3.1 |

*2.5. Application to Real Samples*

The method was used to analyze five PCMs in three tap water and three river water samples from Beijing. The results showed that no PCMs were detected in the tap water from Beijing, while HHCB and tonalide (AHTN) were found in one of the river water samples, with concentrations of 18.7 and 11.7 ng/L for HHCB and AHTN, respectively.

## 3. Materials and Methods

*3.1. Reagents and Materials*

Standard solutions of five polycyclic musks, namely, celestolide (ADBI), phantolide (AHMI), traseolide (ATII), galaxolide (HHCB), and tonalide (AHTN), with concentrations of 10 mg/L were obtained from Dr. Ehrenstorfer (Augsburg, Germany). The internal standard, $^{13}C_6$-labeled hexachlorobenzene ($^{13}C_6$-HCB, 100 mg/L), was obtained from Cambridge Isotope Laboratories (Andover, MA, USA). Pesticide-grade *n*-hexane, dichloromethane (DCM), acetone, and chromatography-grade ethyl acetate were supplied by Fisher Scientific (J.T. Baker, Pittsburgh, PA, USA). *N,N*′-Dicyclohexylcarbodiimide (DCC) was obtained from Alfa Aesar (Ward Hill, MA, USA). Hydrazine hydrate (85%) was purchased from Sinopharm Chemical Reagent Co., Ltd. (Shanghai, China). Analytical-grade *N,N*-dimethylformamide (DMF) was obtained from J&K Scientific Ltd. (Beijing, China).

Graphene oxides (GOs) were purchased from Nanjing XFNANO Materials Tech (Nanjing, China). Aminosilica and C18 silica were obtained from Agilent (Santa Clara, CA, USA). *N*-Vinylpyrrolidone polymeric cartridges (Oasis HLB) were obtained from Waters (Milford, MA, USA).

## 3.2. Preparation and Characterization of rGO@silica

GO@silica hybrids were prepared by linking the GO carboxy groups with amino groups on spherical aminosilica microparticles, similar to the procedure reported by Liu et al. [21]. To summarize, GO (100 mg) was dispersed in DMF (150 mL) by sonication for 30 min. Next, aminosilica (1 g) and DCC (100 mg) were added, and the mixture was stirred at 50 °C for 24 h. The solid product was washed with water and methanol several times to remove unbound GO. The collected GO@silica was dried at 60 °C for 12 h. rGO@silica was obtained by reducing GO@silica with hydrazine. Briefly, GO@silica (1 g) and hydrazine hydrate (0.5 mL, 85%) were added to water (50 mL), and the mixture was heated at 95 °C for 2 h. The solid product was filtered, washed with ultrapure water and methanol, and dried at 60 °C for 12 h.

The surface morphology of the materials was characterized using field-emission scanning electron microscopy (FESEM; Hitachi-S-4800, Hitachi, Tokyo, Japan). X-ray photoelectron spectroscopy (XPS) was performed on a Thermo Scientific ESCALAB 250Xi XPS system (Waltham, MA, USA). The surface functional groups were analyzed by Fourier transform infrared spectroscopy (FT-IR) using a Nicolet 6700 FT-IR spectrometer (Thermo Nicolet, Madison, WI, USA).

## 3.3. Analytical Procedure

SPE was performed on a SPE Vacuum Manifold apparatus (Sigma-Aldrich, St. Louis, MO, USA) equipped with a V700 vacuum pump (Buchi, Flawil, Switzerland). To prepare the SPE cartridges, rGO@silica powder (100 mg) was packed into a 3 mL polypropylene column with an upper and a lower frit to avoid sorbent loss. For PCM enrichment, the cartridge was preconditioned with methanol (10 mL) and water (10 mL). The aqueous sample solution containing 0.5% methanol was then passed through the cartridge at a flow rate of about 5 mL/min. After extraction, the cartridge was washed with a 5% methanol aqueous solution (10 mL) and then was dried under vacuum for 5 min to remove residual water. The retained analytes were then eluted with DCM (8 mL). The effluent was evaporated to dryness under a stream of nitrogen and reconstituted in *n*-hexane (1 mL). Before GC–MS/MS analysis, the extracts were spiked with the internal standard (100 ng).

The GC–MS/MS system consisted of an Agilent 7890B GC instrument equipped with an Agilent 7693B autosampler and an Agilent 7000C triple quadrupole system (Agilent Technologies, Palo Alto, CA, USA). GC separation was performed using two identical Agilent J&W HP—5 ms UI capillary columns (15 m × 0.25 mm I.D., 0.25-μm film thickness) connected through an auxiliary programmable control module. Backflushing was performed to shorten the analysis time and reduce the system maintenance. The injector temperature was held at 280 °C for the entire run. Helium (99.999%) was used as the carrier gas at a constant flow rate of 1 mL/min. The oven temperature program was set as follows: initial temperature, 60 °C for 1 min; increased to 170 °C at 40 °C/min; then increased to 230 °C at 10 °C/min; then ramped to 280 °C at 30 °C/min; and finally held at 280 °C for 1 min. The total run time was 12.4 min, plus an additional 3 min for backflushing at 300 °C.

The multiple reaction monitoring (MRM) mode was used for monitoring and for the confirmation analysis. The manifold temperature was maintained at 230 °C. Quad MS1 and MS2 temperatures were set to 150 °C. The flow rate of the collision gas ($N_2$) was set to 1.5 mL/min. Two MS/MS transitions were used for each analyte, with the sensitivity optimized using collision energy experiments. MS/MS transitions, collision energies, chromatographic retention times, and molecular and chemical information for each PCM are shown in Table S1 (see Supplementary Materials).

## 4. Conclusions

In this study, a simple and novel method was developed for the enrichment and determination of five PCMs in environmental water samples. The rGO@silica was synthesized by graphene oxide and aminosilica as SPE material, showing high adsorption capacity for PCMs. GC–MS/MS was employed for the quantification. Under the optimized condition, satisfactory recoveries, low LODs, and good repeatability were observed for both tap and river water samples. The LOD and LOQ for the five PCMs were 0.3–0.8 ng/L and 1.1–2.1 ng/L, respectively. The recoveries were 86.6–105.9% for tap water and 82.9–107.1% for river water samples, with relative standard deviations <10%. The obtained results indicated that rGO@silica has great potential in the separation and preconcentration of PCMs from water samples. Additionally, this developed method was used to analyze PCMs in tap and river water samples from Beijing, China. HHCB and AHTN were the main PCM components detected, at concentrations of 18.7 ng/L for HHCB, and 11.7 ng/L for AHTN, in one river water sample.

**Supplementary Materials:** The following are available online, Table S1: Chemical information and MS/MS parameters of five PCMs and the internal standard. Figure S1: FT–IR spectra of aminosilica and prepared rGO@silica. Figure S2: Overview of XPS spectra for aminosilica and prepared rGO@silica. Figure S3: Typical chromatograms of tap water spiked with PCMs (100 ng/L): (A) TIC mode; (B) MRM mode.

**Acknowledgments:** This research was funded by the National Natural Science Foundation of China (No. 41401540), the Beijing Natural Science Foundation (8182021), the Youth Scientific Funds of Beijing Academy of Agriculture and Forestry Sciences (No. QNJJ201531), the Beijing Excellent Talent Project (2015000020060G131) and the Construction Project of Science and Technology Innovation Capacity of the Beijing Academy of Agriculture and Forestry Sciences (KJCX20180112).

**Author Contributions:** Anxiang Lu and Jihua Wang conceived and designed the experiments; Jiayi Chen and Yan Chen performed the experiments; Hua Ping and Cheng Li analyzed the data; Cheng Li contributed reagents, materials, analysis tools; Cheng Li wrote the paper.

**Conflicts of Interest:** The authors declare no conflict of interest.

# References

1. Montes-Grajales, D.; Fennix-Agudelo, M.; Miranda-Castro, W. Occurrence of personal care products as emerging chemicals of concern in water resources: A review. *Sci. Total Environ.* **2017**, *595*, 601–614. [CrossRef] [PubMed]
2. Peng, F.-J.; Pan, C.-G.; Zhang, M.; Zhang, N.-S.; Windfeld, R.; Salvito, D.; Selck, H.; Van den Brink, P.J.; Ying, G.-G. Occurrence and ecological risk assessment of emerging organic chemicals in urban rivers: Guangzhou as a case study in china. *Sci. Total Environ.* **2017**, *589*, 46–55. [CrossRef] [PubMed]
3. McDonough, C.A.; Helm, P.A.; Muir, D.; Puggioni, G.; Lohmann, R. Polycyclic musks in the air and water of the lower great lakes: Spatial distribution and volatilization from surface waters. *Environ. Sci. Technol.* **2016**, *50*, 11575–11583. [CrossRef] [PubMed]
4. Rimkus, G.G. Polycyclic musk fragrances in the aquatic environment. *Toxicol. Lett.* **1999**, *111*, 37–56. [CrossRef]
5. Lange, C.; Kuch, B.; Metzger, J.W. Occurrence and fate of synthetic musk fragrances in a small german river. *J. Hazard. Mater.* **2015**, *282*, 34–40. [CrossRef] [PubMed]
6. Wang, X.; Yuan, K.; Liu, H.; Lin, L.; Luan, T. Fully automatic exposed and in-syringe dynamic single-drop microextraction with online agitation for the determination of polycyclic musks in surface waters of the pearl river estuary and south china sea. *J. Sep. Sci.* **2014**, *37*, 1842–1849. [CrossRef] [PubMed]
7. Fromme, H.; Otto, T.; Pilz, K. Polycyclic musk fragrances in different environmental compartments in Berlin (Germany). *Water Res.* **2001**, *35*, 121–128. [CrossRef]
8. Składanowski, A.C.; Stepnowski, P.; Kleszczyński, K.; Dmochowska, B. Amp deaminase in vitro inhibition by xenobiotics: A potential molecular method for risk assessment of synthetic nitro- and polycyclic musks, imidazolium ionic liquids and n-glucopyranosyl ammonium salts. *Environ. Toxicol. Pharmacol.* **2005**, *19*, 291–296. [CrossRef] [PubMed]
9. Chen, C.; Zhou, Q.; Bao, Y.; Li, Y.; Wang, P. Ecotoxicological effects of polycyclic musks and cadmium on seed germination and seedling growth of wheat (*Triticum aestivum*). *J. Environ. Sci.* **2010**, *22*, 1966–1973. [CrossRef]
10. Fang, H.; Gao, Y.; Wang, H.; Yin, H.; Li, G.; An, T. Photo-induced oxidative damage to dissolved free amino acids by the photosensitizer polycyclic musk tonalide: Transformation kinetics and mechanisms. *Water Res.* **2017**, *115*, 339–346. [CrossRef] [PubMed]

11. Dodson, R.E.; Nishioka, M.; Standley, L.J.; Perovich, L.J.; Brody, J.G.; Rudel, R.A. Endocrine disruptors and asthma-associated chemicals in consumer products. *Environ. Health Perspect.* **2012**, *120*, 935–943. [CrossRef] [PubMed]

12. Howard, P.H.; Muir, D.C.G. Identifying new persistent and bioaccumulative organics among chemicals in commerce. *Environ. Sci. Technol.* **2010**, *44*, 2277–2285. [CrossRef] [PubMed]

13. Marchal, M.; Beltran, J. Determination of synthetic musk fragrances. *Int. J. Environ. Anal. Chem.* **2016**, *96*, 1213–1246. [CrossRef]

14. Schüssler, W.; Nitschke, L. Determination of trace amounts of Galaxolide®(HHCB) by HPLC. *Fresenius J. Anal. Chem.* **1998**, *361*, 220–221. [CrossRef]

15. Kuklenyik, Z.; Bryant, X.A.; Needham, L.L.; Calafat, A.M. SPE/SPME–GC/MS approach for measuring musk compounds in serum and breast milk. *J. Chromatogr. B* **2007**, *858*, 177–183. [CrossRef] [PubMed]

16. Wang, H.; Zhang, J.; Gao, F.; Yang, Y.; Duan, H.; Wu, Y.; Berset, J.-D.; Shao, B. Simultaneous analysis of synthetic musks and triclosan in human breast milk by gas chromatography tandem mass spectrometry. *J. Chromatogr. B* **2011**, *879*, 1861–1869. [CrossRef] [PubMed]

17. Lee, I.; Gopalan, A.-I.; Lee, K.-P. Enantioselective determination of polycyclic musks in river and wastewater by GC/MS/MS. *Int. J. Environ. Res. Public Health* **2016**, *13*, 349. [CrossRef] [PubMed]

18. Chen, L.; Zhou, T.; Zhang, Y.; Lu, Y. Rapid determination of trace sulfonamides in fish by graphene-based SPE coupled with UPLC/MS/MS. *Anal. Methods* **2013**, *5*, 4363–4370. [CrossRef]

19. Speltini, A.; Sturini, M.; Maraschi, F.; Consoli, L.; Zeffiro, A.; Profumo, A. Graphene-derivatized silica as an efficient solid-phase extraction sorbent for pre-concentration of fluoroquinolones from water followed by liquid-chromatography fluorescence detection. *J. Chromatogr. A* **2015**, *1379*, 9–15. [CrossRef] [PubMed]

20. Liu, Q.; Shi, J.; Zeng, L.; Wang, T.; Cai, Y.; Jiang, G. Evaluation of graphene as an advantageous adsorbent for solid-phase extraction with chlorophenols as model analytes. *J. Chromatogr. A* **2011**, *1218*, 197–204. [CrossRef] [PubMed]

21. Ye, Q.; Liu, L.H.; Chen, Z.B.; Hong, L.M. Analysis of phthalate acid esters in environmental water by magnetic graphene solid phase extraction coupled with gas chromatography-mass spectrometry. *J. Chromatogr. A* **2014**, *1329*, 24–29. [CrossRef] [PubMed]

22. Han, Q.; Wang, Z.; Xia, J.; Xia, L.; Chen, S.; Zhang, X.; Ding, M. Graphene as an efficient sorbent for the SPE of organochlorine pesticides in water samples coupled with GC–MS. *J. Sep. Sci.* **2013**, *36*, 3586–3591. [CrossRef] [PubMed]

23. Wang, Z.; Han, Q.; Xia, J.; Xia, L.; Ding, M.; Tang, J. Graphene-based solid-phase extraction disk for fast separation and preconcentration of trace polycyclic aromatic hydrocarbons from environmental water samples. *J. Sep. Sci.* **2013**, *36*, 1834–1842. [CrossRef] [PubMed]

24. Wu, X.; Hong, H.; Liu, X.; Guan, W.; Meng, L.; Ye, Y.; Ma, Y. Graphene-dispersive solid-phase extraction of phthalate acid esters from environmental water. *Sci. Total Environ.* **2013**, *444*, 224–230. [CrossRef] [PubMed]

25. Zhang, G.J.; Li, Z.; Zang, X.H.; Wang, C.; Wang, Z. Solid-phase microextraction with a graphene-composite-coated fiber coupled with GC for the determination of some halogenated aromatic hydrocarbons in water samples. *J. Sep. Sci.* **2014**, *37*, 440–446. [CrossRef] [PubMed]

26. Liu, Q.; Shi, J.; Sun, J.; Wang, T.; Zeng, L.; Jiang, G. Graphene and graphene oxide sheets supported on silica as versatile and high-performance adsorbents for solid-phase extraction. *Angew. Chem. Int. Ed.* **2011**, *50*, 5913–5917. [CrossRef] [PubMed]

27. Ye, N.; Shi, P.; Wang, Q.; Li, J. Graphene as solid-phase extraction adsorbent for CZE determination of sulfonamide residues in meat samples. *Chromatographia* **2013**, *76*, 553–557. [CrossRef]

28. Wu, J.; Chen, L.; Mao, P.; Lu, Y.; Wang, H. Determination of chloramphenicol in aquatic products by graphene-based SPE coupled with HPLC-MS/MS. *J. Sep. Sci.* **2012**, *35*, 3586–3592. [CrossRef] [PubMed]

29. Luo, X.; Zhang, F.; Ji, S.; Yang, B.; Liang, X. Graphene nanoplatelets as a highly efficient solid-phase extraction sorbent for determination of phthalate esters in aqueous solution. *Talanta* **2014**, *120*, 71–75. [CrossRef] [PubMed]

30. Maidatsi, K.V.; Chatzimitakos, T.G.; Sakkas, V.A.; Stalikas, C.D. Octyl-modified magnetic graphene as a sorbent for the extraction and simultaneous determination of fragrance allergens, musks, and phthalates in aqueous samples by gas chromatography with mass spectrometry. *J. Sep. Sci.* **2015**, *38*, 3758–3765. [CrossRef] [PubMed]

31. Li, S.; Zhu, F.; Jiang, R.; Ouyang, G. Preparation and evaluation of amino modified graphene solid-phase microextraction fiber and its application to the determination of synthetic musks in water samples. *J. Chromatogr. A* **2016**, *1429*, 1–7. [CrossRef] [PubMed]

32. Luo, Y.-B.; Zhu, G.-T.; Li, X.-S.; Yuan, B.-F.; Feng, Y.-Q. Facile fabrication of reduced graphene oxide-encapsulated silica: A sorbent for solid-phase extraction. *J. Chromatogr. A* **2013**, *1299*, 10–17. [CrossRef] [PubMed]

**Sample Availability:** Not available.

*molecules*

MDPI

*Article*

# Sensitive Detection of 8-Nitroguanine in DNA by Chemical Derivatization Coupled with Online Solid-Phase Extraction LC-MS/MS

Chiung-Wen Hu [1,†], Yuan-Jhe Chang [2,†], Jian-Lian Chen [3], Yu-Wen Hsu [2,4] and Mu-Rong Chao [2,5,*]

[1]  Department of Public Health, Chung Shan Medical University, Taichung 402, Taiwan; windyhu@csmu.edu.tw
[2]  Department of Occupational Safety and Health, Chung Shan Medical University, Taichung 402, Taiwan; handsom1005@gmail.com (Y.-J.C.); yayen0619@gmail.com (Y.-W.H.)
[3]  School of Pharmacy, China Medical University, Taichung 404, Taiwan; cjl@mail.cmu.edu.tw
[4]  Department of Optometry, Da-Yeh University, Changhua 515, Taiwan
[5]  Department of Occupational Medicine, Chung Shan Medical University Hospital, Taichung 402, Taiwan
*  Correspondence: chaomurong@gmail.com or mrchao@csmu.edu.tw;
    Tel.: +886-4-247-300-22 (ext. 12116); Fax: +886-4-232-481-94
†  These authors contributed equally to this work.

Received: 9 February 2018; Accepted: 6 March 2018; Published: 8 March 2018

**Abstract:** 8-Nitroguanine (8-nitroG) is a major mutagenic nucleobase lesion generated by peroxynitrite during inflammation and has been used as a potential biomarker to evaluate inflammation-related carcinogenesis. Here, we present an online solid-phase extraction (SPE) LC-MS/MS method with 6-methoxy-2-naphthyl glyoxal hydrate (MTNG) derivatization for a sensitive and precise measurement of 8-nitroG in DNA. Derivatization optimization revealed that an excess of MTNG is required to achieve complete derivatization in DNA hydrolysates (MTNG: 8-nitroG molar ratio of 3740:1). The use of online SPE effectively avoided ion-source contamination from derivatization reagent by washing away all unreacted MTNG before column chromatography and the ionization process in mass spectrometry. With the use of isotope-labeled internal standard, the detection limit was as low as 0.015 nM. Inter- and intraday imprecision was <5.0%. This method was compared to a previous direct LC-MS/MS method without derivatization. The comparison showed an excellent fit and consistency, suggesting that the present method has satisfactory effectiveness and reliability for 8-nitroG analysis. This method was further applied to determine the 8-nitroG in human urine. 8-NitroG was not detectable using LC-MS/MS with derivatization, whereas a significant false-positive signal was detected without derivatization. It highlights the use of MTNG derivatization in 8-nitroG analysis for increasing the method specificity.

**Keywords:** online solid-phase extraction; LC-MS/MS; peroxynitrite; nitrated DNA lesion; derivatization; isotope-dilution

---

## 1. Introduction

Chronic inflammation has been linked to heart disease, obesity, diabetes and cancer [1,2]. Under chronic inflammatory conditions, exuberant NO production by activated macrophages is believed to be an important tissue-damage mediator as NO can be further converted into several highly reactive species, such as nitrous anhydride, nitrogen dioxide, nitryl chloride and peroxynitrite [3].

Peroxynitrite is a relatively stable reactive species with a half-life of ~one second at physiological pH and can penetrate the nucleus and induce damage in DNA [4,5]. 8-Nitroguanine (8-nitroG) is the first identified peroxynitrite-mediated nitration product. The formation of 8-nitroG is generally rationalized in terms of addition of low reactive •NO$_2$ to the highly oxidizing guanine radical that results from the deprotonation of guanine radical cation initially generated by one-electron oxidation of guanine [6]. 8-NitroG formed in DNA is chemically unstable and can spontaneously depurinate, yielding apurinic sites with the resultant possibility of GC-to-TA mutation [7]. Alternatively, adenine can be preferentially incorporated opposite 8-nitroG during DNA syntheses, resulting in GC-to-TA transversion [8]. Several research groups have focused on the role of 8-nitroG in infection- and inflammation-related carcinogenesis and examined the formation of this lesion in laboratory animals and clinical samples [9,10]. Their studies have shown that the 8-nitroG formation occurred to a much greater extent in cancerous tissue than in the adjacent non-cancerous tissue and that its formation increased with inflammatory grade [11–13], suggesting that 8-nitroG could be a potential biomarker of inflammation-related carcinogenesis [12].

In the past decade, cellular 8-nitroG levels have been largely semi-quantitatively measured by immunohistochemistry [14–16] or quantitatively measured by HPLC with electrochemical detection (ECD) [17,18]. For quantitative measurement, the reported HPLC-ECD methods had comparatively high detection limits of 20–1000 fmol/injection [19,20] and required the reduction of 8-nitroG by a reducing agent (i.e., the reduction of 8-nitroG to 8-aminoguanine) by sodium hydrosulfite [21], which results in low reproducibility owing to the varied reaction efficiencies.

LC-MS/MS has received a great deal of attention in recent years because it can provide a sensitive and selective means for comprehensive measurement of multiple DNA lesions. In our previous work [22], we demonstrated that 8-nitroG is unstable and readily depurinates with a short half-life (e.g., 2.4 h in double-stranded DNA and 1.6 h in single-stranded DNA at 37 °C). We therefore proposed an LC-MS/MS method for the direct measurement of 8-nitroG in DNA and provided a strategy to overcome the chemical instability of 8-nitroG for the quantitative analysis of cellular 8-nitroG. However, this method was hampered by insufficient sensitivity, and the 8-nitroG was not retained well on the reversed-phase columns, decreasing the separation efficiency.

In this study, we describe a chemical derivatization coupled with online solid-phase extraction (SPE) LC-MS/MS analysis for the sensitive determination of 8-nitroG in DNA. To investigate its effectiveness and reliability, the present method was further compared to direct LC-MS/MS measurement without chemical derivatization [22].

## 2. Results

### 2.1. LC-MS/MS Characteristics of 8-NitroG-MTNG

Figure 1 shows an example chromatogram of 8-nitroG-MTNG and its isotope internal standard [$^{13}$C$_2$,$^{15}$N]-8-nitroG-MTNG of a calf thymus DNA hydrolysate that had been treated with 10 μM ONOO$^-$. The retention times of 8-nitroG-MTNG and [$^{13}$C$_2$,$^{15}$N]-8-nitroG-MTNG are concordant. Low background noise from the biological matrix showed the good selectivity of the method. The negative ESI mass spectrum of 8-nitroG-MTNG contained a [M − H]$^-$ precursor ion at $m/z$ 391 and product ions at $m/z$ 363 (quantifier ion, Figure 1A) and $m/z$ 348 (qualifier ion, Figure 1B) due to loss of CO or C$_2$H$_3$O; a precursor ion at $m/z$ 394 and product ions at $m/z$ 366 (quantifier ion, Figure 1C) and $m/z$ 351 (qualifier ion, Figure 1D) characterized the [$^{13}$C$_2$,$^{15}$N]-8-nitroG-MTNG.

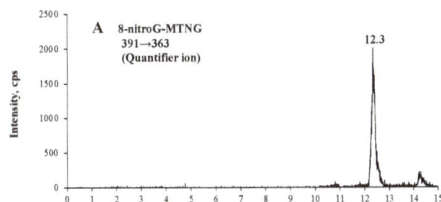

**Figure 1.** *Cont.*

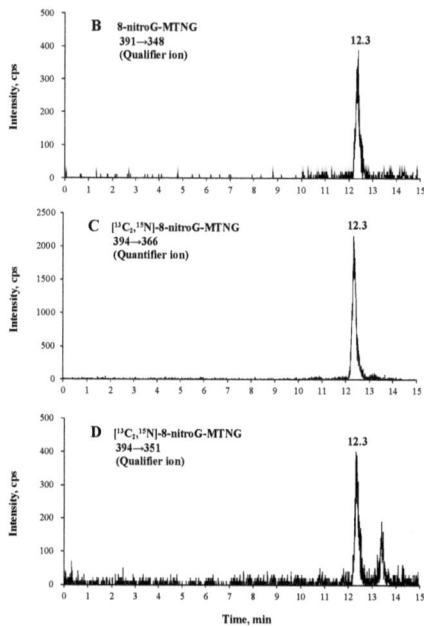

**Figure 1.** Chromatograms of 8-nitroG-MTNG in a hydrolysate of calf thymus DNA that had been treated with 10 μM peroxynitrite, as measured by LC-MS/MS coupled with online SPE. 8-NitroG-MTNG was monitored at $m/z$ 391→363 (**A**) and $m/z$ 391→348 (**B**), and the internal standard [$^{13}C_2$,$^{15}N$]-8-nitroG-MTNG was monitored at $m/z$ 394→366 (**C**) and $m/z$ 394→351 (**D**). cps, counts per second.

## 2.2. Optimization of Derivatization Reaction with MTNG

We investigated the yields for the formation of the conjugate at different molar ratios of MTNG to 8-nitroG (from 232:1 to 14,960:1). As shown in Figure 2, the amounts of 8-nitroG formed increased in a dose-dependent manner with increasing MTNG concentration (0.58–9.35 mM). The derivatization was efficient when the ratio reached 3740:1, where MTNG was added at 9.35 mM. However, we used an even higher MTNG concentration (14.0 mM) for derivatization in this study to ensure a complete derivatization reaction in light of the high variability of biological matrices. This suggested that 5.6 μmole MTNG/μg DNA is needed.

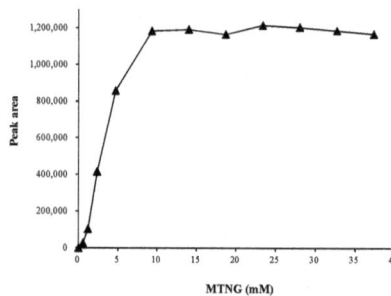

**Figure 2.** Effects of the concentration of added MTNG on the derivatization yield. The peak areas of 8-nitroG-MTNG obtained from the derivatization of a hydrolysate of calf thymus DNA containing 1 μM 8-nitroG. Points denote the mean values of duplicates.

## 2.3. Method Validation

The limit of quantification (LOQ) was defined as the lowest 8-nitroG-MTNG sample concentration meeting prespecified requirements for precision and accuracy within 20%. Using the present method, the LOQ was determined in DNA hydrolysates to be 0.05 nM. The limit of detection (LOD), defined as the lowest concentration that gave a signal-to-noise ratio of at least 3 in DNA hydrolysates, was found to be 0.015 nM (0.3 fmol, see Supplementary Materials, Figure S1), which corresponds to 0.15 μmol 8-nitroG/mol of guanine when using 50 μg of DNA per analysis.

The calibration curve consisted of seven calibration points from 0.16 to 10.3 pmol, and each calibrator contained 0.25 pmol [$^{13}$C$_2$,$^{15}$N]-8-nitroG. The resulting peak area ratios (analyte to internal standard) were plotted against the corresponding pmol (Supplementary Material, Figure S2). Linear regression calculations were unweighted and non-zero-forced, and the regression equation was calculated as $y = 3.9531x + 0.061$. The observed correlation coefficients ($R^2$) during validation were consistently greater than 0.999. All of the calibrators fell within 5% deviation of back-calculated concentrations from nominal spiked concentrations, with an imprecision (CV) < 10%. Meanwhile, the peak identity of 8-nitroG-MTNG in DNA hydrolysates was confirmed by comparing the peak area ratios (quantifier ion/qualifier ion) with those of the calibrators. As an acceptance criterion, ratios in DNA samples should not deviate by more than ±25% from the mean ratios in the calibrators.

The intraday and interday imprecisions was determined from the analysis of three independent DNA samples, which were respectively treated with 50, 100 and 200 μM peroxynitrite. Intraday imprecision was estimated within one batch by analyzing six replicates, whereas interday imprecision was estimated on six separate occasions occurring over a period of 10 days. As shown in Table 1, intraday imprecision was determined to be 1.0–2.7%, and interday imprecision was 2.0–2.5%.

**Table 1.** Precision of isotope-dilution LC-MS/MS method with MTNG derivatization for analysis of 8-nitroG in DNA.

| Characteristics for 8-nitroG [a] | Sample 1 | Sample 2 | Sample 3 |
|---|---|---|---|
| Intraday variation (pmol, mean ± SD) [b] (CV, %) | 0.67 ± 0.02 (2.4) | 0.90 ± 0.02 (2.7) | 1.42 ± 0.01 (1.0) |
| Interday variation (pmol, mean ± SD) [b] (CV, %) | 0.64 ± 0.02 (2.5) | 0.90 ± 0.01 (2.0) | 1.37 ± 0.03 (2.0) |

[a] 50 μL aliquots of 6 μg/mL calf thymus DNA were individually treated with peroxynitrite at three different concentrations (50 μM for sample 1, 100 μM for sample 2 and 200 μM for sample 3); [b] Each DNA solution was analyzed 6 times for the intraday and interday tests; the interday test was performed over a period of 10 days.

Recovery was evaluated by adding unlabeled standard mixture at five different levels to three peroxynitrite-treated DNA samples and measuring three replicates of these samples. As shown in (Supplementary Material, Figure S3), the mean recoveries were 96–104%, 99–105% and 96–104%, respectively, for those three DNA samples as estimated from the increase in the measured amount after addition of the analyte divided by the amount added, while the recoveries as calculated from the slope of the regression were 103%, 98% and 103% ($R^2 > 0.99$), respectively.

Matrix effects were calculated according to the following equation:

$$\text{Matrix effects} = 1 - \left( \frac{\text{Peak area of internal standard in the presence of matrix}}{\text{Peak area of internal standard in the absence of matrix}} \right) \times 100\% \quad (1)$$

The peak area in the presence of matrix refers to the peak area of the internal standard in DNA samples, while the peak area in the absence of matrix refers to the peak area of the internal standard prepared in deionized water. The relative change in the peak area of the internal standard was attributed to matrix effects, which reflect the combination of reduced derivatization efficiency, online extraction loss and ion suppression due to the DNA matrix. In this study, the matrix effects for 8-nitroG-MTNG were less than 20% in all DNA samples. Although the use of stable isotope-labeled internal standards could have compensated for different matrix effects, the low matrix effect achieved in this study ensures the high sensitivity of the method.

### 2.4. 8-NitroG in Calf Thymus DNA Treated with Peroxynitrite

The present method was applied to quantify the levels of 8-nitroG in calf thymus DNA, which was incubated with various concentrations (2.5–200 µM) of peroxynitrite. 8-NitroG was not detected in control DNA. In the peroxynitrite-treated samples, the levels of 8-nitroG increased in a dose-dependent manner with peroxynitrite concentration (Supplementary Material, Figure S4). 8-NitroG was formed at a level of 211 µmol/mol of guanine even with a peroxynitrite concentration as low as 2.5 µM. The formation of 8-nitroG reached a maximum level of 5823 µmol/mol of guanine when treated with 200 µM peroxynitrite.

### 2.5. Comparison between 8-NitroG Analysis Using Online SPE LC-MS/MS Method with and without MTNG Derivatization

8-NitroG concentrations determined using the proposed online SPE LC-MS/MS following MTNG derivatization were compared to concentrations derived from the same samples using a direct online LC-MS/MS method without derivatization. As shown in Figure 3, regression analysis showed that the two methods were highly correlated (Pearson $R^2 = 0.9893$, $p < 0.001$, $n = 39$). The 8-nitroG levels derived from the present method were close to those obtained from the reported direct measurement by online SPE LC-MS/MS without derivatization, giving a slope of 1.01.

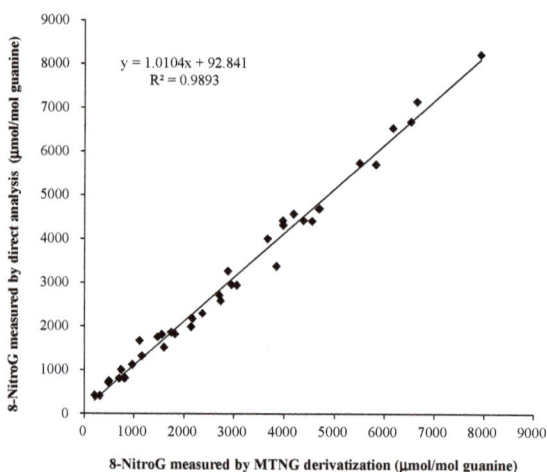

**Figure 3.** Correlation between quantitative results obtained through online SPE LC-MS/MS analyses with (this work) and without glyoxal derivatization [22]. The DNA samples containing various levels of 8-nitroG were prepared by treating calf thymus DNA with peroxynitrite at concentrations of 2.5–300 µM.

## 3. Discussion

Despite numerous attempts, unambiguous evidence for the formation of 8-nitroG in cellular DNA or animal organs has not yet been provided. Apparently, the presence of 8-nitroG was detected by immunohistochemical methods in inflamed tissues [13]. However, the specificity of monoclonal/polyclonal antibodies against single oxidized base (such as 8-oxo-7,8-dihydroguanine) has been questioned due to the occurrence of cross-reactivity with overwhelming guanine bases [23]. The data provided by the several HPLC-ECD methods [19–21] that were aimed at measuring 8-nitroG, mostly as amino derivatives (8-aminoguaine), might not also be convincing. Therefore, there is a strong need of more accurate methods such as LC-MS/MS that appears to be the gold standard analytical tool for detecting DNA base lesions.

Our method coupling LC-MS/MS with derivatization and stable isotope-dilution facilitates the accurate and sensitive detection of cellular 8-nitroG. We further performed derivatization optimization, which is essential for an accurate measurement. The results revealed that an excess of MTNG is required (MTNG:8-nitroG molar ratio 3740:1, Figure 2) to completely derivatize 8-nitroG in DNA hydrolysates. This may be attributable to the fact that MTNG reacts not only with 8-nitroG but also with other guanine compounds present in biological samples [24].

Our method to estimate the cellular levels of 8-nitroG has some notable benefits compared with previously reported methods. The primary feature of our method is its high sensitivity (LOD: 0.015 nM), which may allow for the detection of extremely low levels of 8-nitroG in cellular DNA or urine. As shown in Supplementary Material, Figure S5, the sensitivity of the present method increased greatly with the derivatization, by approximately 10 times, compared to the method without derivatization (direct measurement [22]). Previously, Villaño et al. [25] and Wu et al. [26] developed HPLC-MS/MS methods for direct determination of 8-nitroG in plasma and urine, respectively, and reported LODs of 0.15–0.4 nM. One previous study by Ishii et al. [27] also attempted to measure 8-nitroG using LC-MS following MTNG derivatization. However, their method was not validated with a LOD (~1 nM) approximately 70 times higher than our method and a low specificity due to only the precursor ion was monitored. Meanwhile, it is also noted that that the MTNG amount used in the work of Ishii et al. [27] was significantly insufficient (0.09 μmole/μg DNA) for a complete derivatization reaction as compared to our finding (5.6 μmole MTNG/μg DNA). Furthermore, when comparing the LODs of previously reported chromatographic methods (except for LC-MS methods) for 8-nitroG, LODs of 2–100 nM were reported [17,20,28], and those are 100–6000 times higher than our LOD of 0.015 nM. Meanwhile, we have attempted to measure the background level of 8-nitroG in cellular DNA using the proposed method. The cells (human endothelial hybrid cells and Chinese hamster ovary cells) were lysed, subjected to acid hydrolysis [22], and derivatization with MTNG as described above. The 8-nitroG in cellular DNA was found to be non-detectable (see Supplementary Material Figure S6). It was estimated that background level of 8-nitroG in cellular DNA was less than 0.15 μmol/mol of guanine when using 50 μg of DNA per analysis.

The second important feature of our method is the use of the isotope-dilution method for the 8-nitroG measurement, which permits high precision and accuracy. We added the stable isotope-labeled standard to the nucleoside mixtures; thus, the analytes and the corresponding internal standard were derivatized simultaneously with MTNG. Any variation/alteration in experimental conditions (e.g., derivatization efficiency, matrix effect and MS/MS performance variation) will thus be compensated for, and the accuracy of measurement will be ensured. Additionally, the co-elution of the analyte and its isotope-labeled standard along with the similar fragmentation pattern of the analyte and internal standard offers unequivocal chemical specificity for analyte identification [29]. A high accuracy measurement can also be supported by the observation of the high consistency (with a slope of 1.01, Figure 3) of the measured results with and without derivatization. The same strategy was also reported previously by Ishii et al. [27], who employed the stable isotope-labeled internal standard ($[^{13}C_2,^{15}N]$-8-nitroG) in the LC-MS method for 8-nitroG measurement.

The third advantage of the proposed method is the use of online SPE, which avoids ion-source contamination and significant matrix effects, resulting from the presence of a large amount of MTNG. To achieve complete derivatization, an excess of MTNG is required. However, part of this large excess could have remained unreacted and later injected into the LC-MS/MS system. As shown in Figure 4, the above problem was effectively avoided by the use of online SPE; since all unreacted MTNG was washed away during online SPE prior to analytical column chromatography and the ionization process in mass spectrometry. It was estimated that less than 0.01% of unreacted MTNG was introduced to the analytical column (Figure 4B).

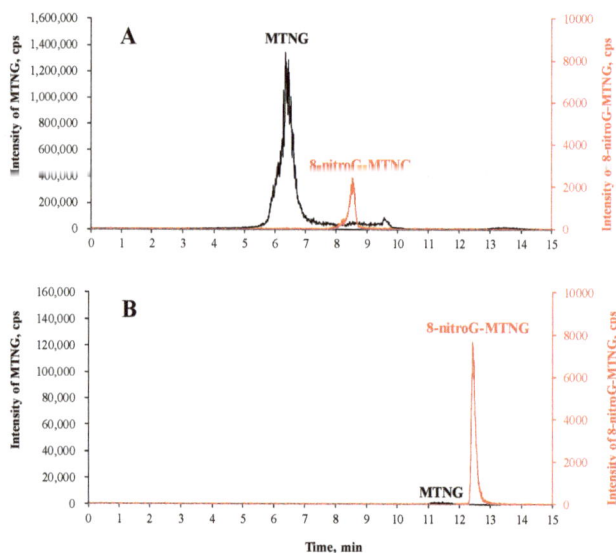

**Figure 4.** Total ion chromatogram of a derivatized DNA sample during trap column separation (**A**) and analytical column separation after column switching (**B**). MTNG was monitored at $m/z$ 215→144 in positive ionization mode with a retention time at 6.3 min and 8-nitroG-MTNG were clearly well separated on the SPE column (**A**), and only the fraction containing 8-nitroG-MTNG at the retention time from 7.5 to 9.5 min was eluted into the analytical column (**B**).

Sawa et al. [30] measured 8-nitroG in urine using HPLC-ECD following immunoaffinity purification. They suggested that 8-nitroG in urine may be a potential biomarker of nitrative damage, although its level in urine could be as low as ~0.01 nM with a low detection rate. Interestingly, several recent studies from the same group have measured significantly high levels of 8-nitroG in the urine of healthy subjects (~4–3700 nM, [26,31,32]) as measured by LC-MS/MS. We presumed that these conflicting results in the literature could be attributed to the interference present in the urine samples as detected by a low-resolution mass spectrometer. To test our assumption, a serial measurement was conducted in 10 urine samples of healthy subjects; these samples were simultaneously measured by three different methods, including LC-MS/MS without derivatization, the present LC-MS/MS method with derivatization and UPLC-high-resolution MS (HRMS) (LTQ-Orbitrap Elite MS, Thermo Fisher Scientific). The UPLC gradient, column material and HRMS parameters applied is provided in Supplementary Material Table S1. A typical comparison of chromatograms of 8-nitroG in urine as measured by the three methods is given (Supplementary Material, Figure S7). A significant signal in a urine sample is noted to have the same transition $m/z$ 195→178 as 8-nitroG and co-eluted with its internal standard (Supplementary Material, Figure S7A) as measured by LC-MS/MS without derivatization. However, when the same urine sample was measured by the present LC-MS/MS method with derivatization or UPLC-HRMS, no 8-nitroG was detected (Supplementary Material, Figure S7B,C). These findings proved that the urinary 8-nitroG signal detected by LC-MS/MS without derivatization (Supplementary Material, Figure S7A) was an interferent. In fact, the urinary 8-nitroG levels could be less than 0.01 nM in healthy subjects (<LOD of Supplementary Material, Figure S7B in urine). Meanwhile, it is worth noting that no 8-nitroG detected in urine by the present LC-MS/MS method with MTNG derivatization (Supplementary Material, Figure S7B) highlights the use of MTNG derivatization in 8-nitroG analysis for increasing the method specificity.

In conclusion, this study describes a sensitive and reliable LC-MS/MS method to quantitatively analyze 8-nitroG in DNA hydrolysates. With the combination of online SPE and MTNG derivatization, matrix interferences are significantly reduced, and the sensitivity of the present method has been highly increased. The present method was compared to a previous direct LC-MS/MS method without chemical derivatization. The comparison showed an excellent fit ($R^2 = 0.9893$, $p < 0.001$) and consistency (slope = 1.01), suggesting that the present method has satisfactory effectiveness and reliability for 8-nitroG analysis. More importantly, an excellent consistency proved that no artifact was produced during our derivatization that frequently encountered in the chemical derivatization of modified DNA bases [33]. Since 8-nitroG presents at trace levels in cells, our method could be useful in both laboratory and clinical research to understand the correlation between inflammation-related DNA damage and carcinogenesis. Subsequently, there is a limitation in the present work, which is the lack of other parameters optimization for MTNG derivatization, including the reaction buffers, reaction time, temperature, pH, etc.

## 4. Materials and Methods

### 4.1. Chemicals

8-NitroG (>98% purity) and $[^{13}C_2, ^{15}N]$-8-nitroguanine ($[^{13}C_2, ^{15}N]$-8-nitroG, >50% purity) were purchased from Toronto Research Chemicals (North York, Ontario, Canada). The purity and concentration of $[^{13}C_2, ^{15}N]$-8-nitroG was quantified by HPLC-UV using unlabeled 8-nitroG standards and confirmed by LC-MS/MS analysis. 6-Methoxy-2-naphthyl glyoxal hydrate (MTNG), diethylenetriaminepentaacetic acid (DTPA), peroxynitrite ($ONOO^-$), calf thymus DNA, acetonitrile (ACN), dimethyl sulfoxide (DMSO) and ammonium acetate (AA) were obtained from Sigma-Aldrich (St. Louis, MO, USA).

### 4.2. Stock and Working Solutions

Standard stock solutions of 8-nitroG and $[^{13}C_2, ^{15}N]$-8-nitroG were individually prepared in 5% ($v/v$) methanol at a concentration of 0.1 mM and stored at $-20\,^\circ$C. A series of standard working solutions of 8-nitroG (3.2–204 nM) were prepared by serial dilution of the stock solution with deionized water. The internal standard solution of $[^{13}C_2, ^{15}N]$-8-nitroG at a concentration of 5 nM was made by diluting the stock solution with deionized water. The MTNG stock solution was initially prepared by dissolving it in DMSO to a final concentration of 50 mM, after which it was protected from light and stored at $-20\,^\circ$C; it was diluted to the desired concentration with DMSO before use.

### 4.3. Nitration of Calf Thymus DNA by Peroxynitrite

The peroxynitrite ($ONOO^-$) solution was carefully thawed and kept on ice. An aliquot of the stock solution was diluted 40-fold with 0.3 N NaOH, and the absorbance at 302 nm was measured with 0.3 N NaOH as blank. The peroxynitrite concentration was calculated using a molar absorption coefficient of 1670 $M^{-1}$ $cm^{-1}$. Fifty microliters of the peroxynitrite prepared in 0.3 N NaOH at various concentrations was added to 150 µL of reaction mixture that contained 50 µL of 6 µg/mL calf thymus DNA, 50 µL of 1 M AA (pH 7.4) and 50 µL of 1 mM DTPA (a metal chelator) in 0.3 N HCl. The sample was mixed by vortexing for 1 min at room temperature with a final pH at ~7.4. As the half-life of peroxynitrite is only 1–2 s near neutral pH, the reaction completed rapidly. Control experiments were performed using decomposed $ONOO^-$ in NaOH, obtained by leaving the peroxynitrite solution overnight at room temperature, after which $ONOO^-$ was completely decomposed as determined spectrophotometrically [34].

### 4.4. Hydrolysis and Derivatization of DNA Samples for 8-NitroG Analysis

Fifty-microliter DNA samples were spiked with 50 µL of 5 nM $[^{13}C_2, ^{15}N]$-8-nitroG and subjected to acid hydrolysis in 100 µL of 1 N HCl for 30 min at 80 $^\circ$C, followed by neutralization with 100 µL of 1 N NaOH. 8-NitroG derivatization and reaction buffer used were described previously [24,27,35] and was performed with some modifications. Reaction buffer was prepared by combining 10 mL of

20 mM sodium acetate buffer (pH 4.8), 1 mL of 3 M sodium acetate buffer (pH 5.1) and 1 mL of 1 M Tris-HCl buffer (pH 8.0), aliquoted and stored at −20 °C. The samples were derivatized by mixing a portion of DNA hydrolysate (150 µL) with 60 µL of 14 mM MTNG, 150 µL of reaction buffer and 15 µL of 1 N HCl for 90 min at 25 °C to yield 8-nitroG-MTNG. The derivatized sample was transferred to a vial for online SPE LC-MS/MS determination. The chemical structures of 8-nitroG-MTNG and its corresponding internal standard ($[^{13}C_2,^{15}N]$ 8-nitroG-MTNG) are shown in Figure 5. To establish a linear calibration curve, 8-nitroG standards (3.2, 6.4, 12.7, 25.5, 51, 102 and 204 nM) in 50 µL of 6 µg/mL blank (untreated) DNA were mixed with 50 µL of 5 nM $[^{13}C_2,^{15}N]$-8-nitroG and then hydrolyzed and derivatized as described above. The levels of 8-nitroG in DNA were expressed as µmol/mol of guanine. The analysis of guanine was performed by an isotope-dilution LC-MS/MS method previously described by Chao et al. [36].

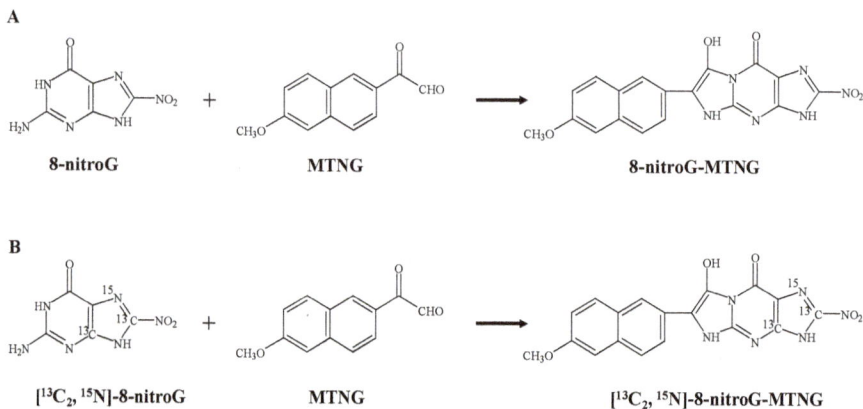

**A**

8-nitroG           MTNG           8-nitroG-MTNG

**B**

$[^{13}C_2, ^{15}N]$-8-nitroG       MTNG       $[^{13}C_2, ^{15}N]$-8-nitroG-MTNG

**Figure 5.** Derivatization of unlabeled 8-nitroG and $[^{13}C_2,^{15}N]$-8-nitroG with MTNG to form 8-nitroG-MTNG (**A**) and $[^{13}C_2,^{15}N]$-8-nitroG-MTNG (**B**).

*4.5. Automated Online Extraction System and Liquid Chromatography*

Column switching was controlled by a multi-channel valve (6-port, 2-position valve, VICI Valco, Houston, TX, USA) according to the pattern shown in detail in a previous publication [37]. An Agilent 1100 series HPLC system (Agilent Technologies, Wilmington, DE, USA) equipped with two binary pumps was used. The detailed column-switching operation sequence is summarized in Table 2. For online purification, an SPE column (33 × 2.1 mm i.d., 5 µm, Inertsil, ODS-3) was employed, while a reversed-phase C18 column (75 × 2.1 mm i.d., 5 µm, Inertsil, ODS-3) was used as the analytical column. The injection volume for the prepared DNA samples was 20 µL. After injection, the SPE column was loaded and washed for 7.5 min with Eluent I at a flow rate of 200 µL/min. After valve switching, 8-nitroG-MTNG were eluted to the analytical column with Eluent II at a flow rate of 200 µL/min. The valve was switched back to the starting position at 9.5 min; the SPE column was then reconditioned for the next run.

**Table 2.** Timetable for the column-switching procedure.

| Time (min) | Eluent I (SPE Column) | | Eluent II (Analytical Column) | | Valve Position | Flow Rate (µL/min) | Remarks |
|---|---|---|---|---|---|---|---|
| | Solvent Ia [a] (%) | Solvent Ib [b] (%) | Solvent IIa [a] (%) | Solvent IIb [b] (%) | | | |
| 0.0 | 70 | 30 | 50 | 50 | A | 200 | Sample injection and washing |
| 7.5 | 70 | 30 | 50 | 50 | B | 200 | Start of elution of 8-nitroG-MTNG to the analytical column |
| 9.5 | 70 | 30 | 50 | 50 | A | 200 | End of elution; SPE column cleanup and reconditioning |
| 10.0 | 70 | 30 | 50 | 50 | A | 200 | |
| 10.1 | 0 | 100 | 50 | 50 | A | 200 | |
| 10.5 | 0 | 100 | 0 | 100 | A | 200 | |
| 11.5 | 0 | 100 | 0 | 100 | A | 200 | |
| 12.0 | 70 | 30 | 50 | 50 | A | 200 | |
| 15.0 | 70 | 30 | 50 | 50 | A | 200 | |

[a] 5% (*v*/*v*) ACN containing 1 mM AA; [b] 80% (*v*/*v*) ACN containing 1 mM AA.

*4.6. ESI-MS/MS*

Mass spectrometric analysis was performed on an API 4000 QTrap hybrid triple quadrupole linear ion trap mass spectrometer (Applied Biosystems, Framingham, MA, USA) equipped with a TurboIonSpray (TIS) source. The resolution was set to a peak width (FWHM) of 0.7 Th for both Q1 and Q3 quadrupoles. Instrument parameters were optimized by infusion experiments with standard derivatives (8-nitroG-MTNG and [$^{13}$C$_2$,$^{15}$N]-8-nitroG-MTNG) in negative ionization mode. Prior to infusion, the standard derivatives were purified by a manual C18 SPE to remove the salts; the standard derivatives were loaded onto a Sep-Pak C18 cartridge (100 mg/1 mL, Waters, Milford, MA, USA) preconditioned with methanol and deionized water. The cartridge was then washed with 1 mL of 20% methanol and eluted with 1 mL of 60% methanol. The eluate was suitable for precursor and product ion scan. Detailed product ion spectra of 8-nitroG-MTNG and its corresponding internal standard ([$^{13}$C$_2$,$^{15}$N]-8-nitroG-MTNG) are given in Figure 6.

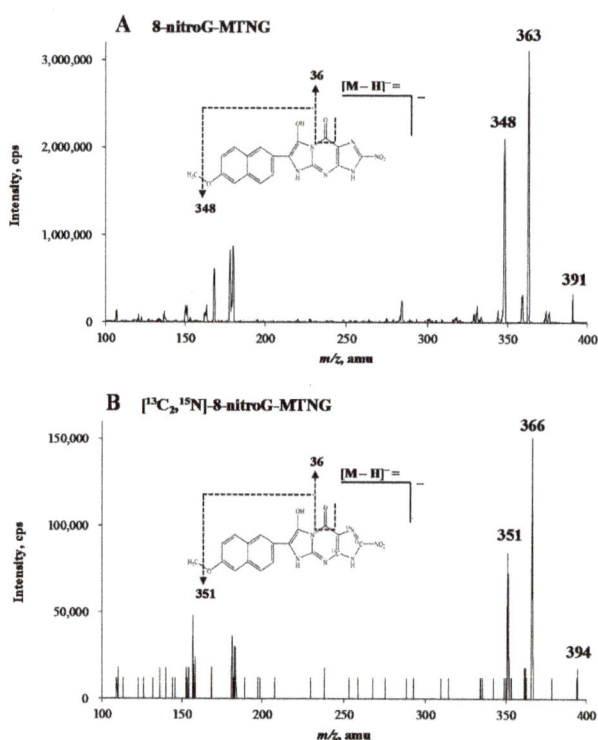

**Figure 6.** Negative ion electrospray MS/MS spectra of [M − H]⁻ of 8-nitroG-MTNG (**A**) and [$^{13}$C$_2$,$^{15}$N]-8-nitroG-MTNG (**B**).

The ion spray voltage was maintained at −4500 V. The TIS source temperature was set at 450 °C. Ion source gas 1 (GS1) was set at 70 (arbitrary unit), ion source gas 2 (GS2) at 70, curtain gas at 10, and collision-activated dissociation gas at medium. Detection was performed in multiple reaction monitoring (MRM) mode. The precursor and product ions, along with optimized parameters, are given in Table 3. The most abundant fragment ion was used for quantification (quantifier ion), and the second most abundant ion was used for qualification (qualifier ion). Analyst 1.4.2 software (Applied Biosystems) was used for data acquisition and processing.

*Molecules* **2018**, 23, 605

**Table 3.** Tandem mass spectrometry parameters for 8-nitroG-MTNG and [$^{13}C_2,^{15}N$]-8-nitroG-MTNG.

| Compound | Q1 Mass (amu) | Q3 Mass (amu) | Dwell Time (ms) | DP [a] (V) | EP [b] (V) | CXP [c] (V) | CE [d] (V) |
|---|---|---|---|---|---|---|---|
| 8-nitroG-MTNG | 391 | 363 [e] | 100 | −50 | −11 | −11 | −30 |
| | 391 | 348 | 100 | −50 | −11 | −11 | −40 |
| [$^{13}C_2,^{15}N$]-8-nitroG-MTNG | 394 | 366 [e] | 100 | −50 | −11 | −11 | −30 |
| | 394 | 351 | 100 | −50 | −11 | −11 | −45 |

[a] Declustering potential; [b] Entrance potential; [c] Collision cell exit potential; [d] Collision energy; [e] Quantifier transition.

### 4.7. Optimization of MTNG Derivatization

The optimal amount of MTNG addition for derivatization was first investigated. The optimization test was performed by mixing 150 μL of DNA hydrolysate containing 1 μM 8-nitroG with 150 μL of reaction buffer, 15 μL 1 N HCl and 60 μL of various concentrations of MTNG (0.58–37.4 mM). The resulting mixture was then incubated at 25 °C for 90 min, followed by online SPE LC-MS/MS analysis.

### 4.8. Direct Measurement of 8-NitroG by Online SPE LC-MS/MS without Derivatization

The 8-nitroG levels in peroxynitrite-treated DNA were measured in parallel by a recently reported online SPE LC-MS/MS method without derivatization [22]. Briefly, the treated DNA samples (50 μL) were spiked with 50 μL of 10 ng/mL [$^{13}C_2,^{15}N$]-8-nitroG, subjected to acid hydrolysis by adding 100 μL of 1 N HCl for 30 min at 80 °C, neutralized with 100 μL of 1 N NaOH and directly analyzed by online SPE LC-MS/MS. The samples were analyzed in the negative ion MRM mode. The 8-nitroG was monitored at $m/z$ 195→178 (quantifier ion) and 195→153 (qualifier ion), and [$^{13}C_2,^{15}N$]-8-nitroG was monitored at $m/z$ 198→181.

**Supplementary Materials:** The Supplementary Materials are available online.

**Acknowledgments:** The authors acknowledge financial support from the Ministry of Science and Technology, Taiwan [grant numbers NSC 102-2314-B-040-016-MY3 and MOST 105-2314-B-040-005]. The authors thank Jia-Hong Lin, Cheng-Cheng Lin and Chih-Hung Hu for help with sample preparation and measurements.

**Author Contributions:** Mu-Rong Chao and Chiung-Wen Hu conceived and designed the experiments; Yuan-Jhe Chang and Yu-Wen Hsu performed the experiments; Mu-Rong Chao, Jian-Lian Chen and Chiung-Wen Hu analyzed the data; Chiung-Wen Hu and Yuan-Jhe Chang wrote the paper.

**Conflicts of Interest:** The authors declare no conflict of interest.

## Abbreviations

| | |
|---|---|
| 8-nitroG | 8-Nitroguanine |
| LC-MS/MS | liquid chromatography-tandem mass spectrometry |
| LOD | limit of detection |
| LOQ | imit of quantification |
| MTNG | 6-methoxy-2-naphthyl glyoxal hydrate |
| MRM | multiple reaction monitoring |
| ONOO⁻ | peroxynitrite |
| SPE | solid-phase extraction |
| UPLC-HRMS | ultra-performance liquid chromatography-high resolution mass spectrometry |

## References

1. Fougere, B.; Boulanger, E.; Nourhashemi, F.; Guyonnet, S.; Cesari, M. Chronic Inflammation: Accelerator of Biological Aging. *J. Gerontol. A Biol. Sci. Med. Sci.* **2016**, 72, 1218–1225. [CrossRef] [PubMed]
2. El Assar, M.; Angulo, J.; Rodriguez-Manas, L. Oxidative stress and vascular inflammation in aging. *Free Radic. Biol. Med.* **2013**, 65, 380–401. [CrossRef] [PubMed]

3. Niles, J.C.; Wishnok, J.S.; Tannenbaum, S.R. Peroxynitrite-induced oxidation and nitration products of guanine and 8-oxoguanine: Structures and mechanisms of product formation. *Nitric Oxide* **2006**, *14*, 109–121. [CrossRef] [PubMed]

4. Beckman, J.S.; Chen, J.; Ischiropoulos, H.; Crow, J.P. Oxidative chemistry of peroxynitrite. *Methods Enzymol.* **1994**, *233*, 229–240. [PubMed]

5. Squadrito, G.L.; Pryor, W.A. Oxidative chemistry of nitric oxide: The roles of superoxide, peroxynitrite, and carbon dioxide. *Free Radic. Biol. Med.* **1998**, *25*, 392–403. [CrossRef]

6. Cadet, J.; Wagner, J.R.; Shafirovich, V.; Geacintov, N.E. One-electron oxidation reactions of purine and pyrimidine bases in cellular DNA. *Int. J. Radiat. Biol.* **2014**, *90*, 423–432. [CrossRef] [PubMed]

7. Ohshima, H.; Sawa, T.; Akaike, T. 8-nitroguanine, a product of nitrative DNA damage caused by reactive nitrogen species: Formation, occurrence, and implications in inflammation and carcinogenesis. *Antioxid. Redox Signal.* **2006**, *8*, 1033–1045. [CrossRef] [PubMed]

8. Suzuki, N.; Yasui, M.; Geacintov, N.E.; Shafirovich, V.; Shibutani, S. Miscoding events during DNA synthesis past the nitration-damaged base 8-nitroguanine. *Biochemistry* **2005**, *44*, 9238–9245. [CrossRef] [PubMed]

9. Hiraku, Y. Oxidative and nitrative DNA damage induced by environmental factors and cancer risk assessment. *Fukuoka Igaku Zasshi* **2014**, *105*, 33–41. [PubMed]

10. Sawa, T.; Ohshima, H. Nitrative DNA damage in inflammation and its possible role in carcinogenesis. *Nitric Oxide* **2006**, *14*, 91–100. [CrossRef] [PubMed]

11. Kawanishi, S.; Hiraku, Y. Oxidative and nitrative DNA damage as biomarker for carcinogenesis with special reference to inflammation. *Antioxid. Redox Signal.* **2006**, *8*, 1047–1058. [CrossRef] [PubMed]

12. Kawanishi, S.; Ohnishi, S.; Ma, N.; Hiraku, Y.; Oikawa, S.; Murata, M. Nitrative and oxidative DNA damage in infection-related carcinogenesis in relation to cancer stem cells. *Genes Environ.* **2016**, *38*, 1–12. [CrossRef] [PubMed]

13. Murata, M.; Thanan, R.; Ma, N.; Kawanishi, S. Role of nitrative and oxidative DNA damage in inflammation-related carcinogenesis. *J. Biomed. Biotechnol.* **2012**, *2012*, 1–11. [CrossRef] [PubMed]

14. Hiraku, Y.; Sakai, K.; Shibata, E.; Kamijima, M.; Hisanaga, N.; Ma, N.; Kawanishi, S.; Murata, M. Formation of the nitrative DNA lesion 8-nitroguanine is associated with asbestos contents in human lung tissues: A pilot study. *J. Occup. Health* **2014**, *56*, 186–196. [CrossRef] [PubMed]

15. Saigusa, S.; Araki, T.; Tanaka, K.; Hashimoto, K.; Okita, Y.; Fujikawa, H.; Okugawa, Y.; Toiyama, Y.; Inoue, Y.; Uchida, K.; et al. Identification of patients with developing ulcerative colitis-associated neoplasia by nitrative DNA damage marker 8-nitroguanin expression in rectal mucosa. *J. Clin. Gastroenterol.* **2013**, *47*, e80–e86. [CrossRef] [PubMed]

16. Kawanishi, S.; Hiraku, Y.; Pinlaor, S.; Ma, N. Oxidative and nitrative DNA damage in animals and patients with inflammatory diseases in relation to inflammation-related carcinogenesis. *Biol. Chem.* **2006**, *387*, 365–372. [CrossRef] [PubMed]

17. Hsieh, Y.S.; Chen, B.C.; Shiow, S.J.; Wang, H.C.; Hsu, J.D.; Wang, C.J. Formation of 8-nitroguanine in tobacco cigarette smokers and in tobacco smoke-exposed Wistar rats. *Chem. Biol. Interact.* **2002**, *140*, 67–80. [CrossRef]

18. Yermilov, V.; Rubio, J.; Ohshima, H. Formation of 8-nitroguanine in DNA treated with peroxynitrite in vitro and its rapid removal from DNA by depurination. *FEBS Lett.* **1995**, *376*, 207–210. [CrossRef]

19. Chang, H.R.; Lai, C.C.; Lian, J.D.; Lin, C.C.; Wang, C.J. Formation of 8-nitroguanine in blood of patients with inflammatory gouty arthritis. *Clin. Chim. Acta* **2005**, *362*, 170–175. [CrossRef] [PubMed]

20. Ohshima, H.; Yoshie, Y.; Auriol, S.; Gilibert, I. Antioxidant and pro-oxidant actions of flavonoids: Effects on DNA damage induced by nitric oxide, peroxynitrite and nitroxyl anion. *Free Radic. Biol. Med.* **1998**, *25*, 1057–1065. [CrossRef]

21. Tuo, J.; Liu, L.; Poulsen, H.E.; Weimann, A.; Svendsen, O.; Loft, S. Importance of guanine nitration and hydroxylation in DNA in vitro and in vivo. *Free Radic. Biol. Med.* **2000**, *29*, 147–155. [CrossRef]

22. Hu, C.W.; Chang, Y.J.; Hsu, Y.W.; Chen, J.L.; Wang, T.S.; Chao, M.R. Comprehensive analysis of the formation and stability of peroxynitrite-derived 8-nitroguanine by LC-MS/MS: Strategy for the quantitative analysis of cellular 8-nitroguanine. *Free Radic. Biol. Med.* **2016**, *101*, 348–355. [CrossRef] [PubMed]

23. Garratt, L.W.; Mistry, V.; Singh, R.; Sandhu, J.K.; Sheil, B.; Cooke, M.S.; Sly, P.D. Arestcf, Interpretation of urinary 8-oxo-7,8-dihydro-2′-deoxyguanosine is adversely affected by methodological inaccuracies when using a commercial ELISA. *Free Radic. Biol. Med.* **2010**, *48*, 1460–1464. [CrossRef] [PubMed]

24. Katayama, M.; Matsuda, Y.; Kobayashi, K.; Kaneko, S.; Ishikawa, H. Monitoring of 8-oxo-7,8-dihydro-2'-deoxyguanosine in urine by high-performance liquid chromatography after pre-column derivatization with glyoxal reagents. *Biomed. Chromatogr.* **2006**, *20*, 800–805. [CrossRef] [PubMed]
25. Villaño, D.; Vilaplana, C.; Medina, S.; Cejuela-Anta, R.; Martínez-Sanz, J.M.; Gil, P.; Genieser, H.G.; Ferreres, F.; Gil-Izquierdo, A. Effect of elite physical exercise by triathletes on seven catabolites of DNA oxidation. *Free Radic. Res.* **2015**, *49*, 973–983. [CrossRef] [PubMed]
26. Wu, C.; Chen, S.T.; Peng, K.H.; Cheng, T.J.; Wu, K.Y. Concurrent quantification of multiple biomarkers indicative of oxidative stress status using liquid chromatography-tandem mass spectrometry. *Anal. Biochem.* **2016**, *512*, 26–35. [CrossRef] [PubMed]
27. Ishii, Y.; Ogara, A.; Okamura, T.; Umemura, T.; Nishikawa, A.; Iwasaki, Y.; Ito, R.; Saito, K.; Hirose, M.; Nakazawa, H. Development of quantitative analysis of 8-nitroguanine concomitant with 8-hydroxydeoxyguanosine formation by liquid chromatography with mass spectrometry and glyoxal derivatization. *J. Pharm. Biomed. Anal.* **2007**, *43*, 1737–1743. [CrossRef] [PubMed]
28. Li, M.J.; Zhang, J.B.; Li, W.L.; Chu, Q.C.; Ye, J.N. Capillary electrophoretic determination of DNA damage markers: Content of 8-hydroxy-2'-deoxyguanosine and 8-nitroguanine in urine. *J. Chromatogr. B Anal. Technol. Biomed. Life Sci.* **2011**, *879*, 3818–3822. [CrossRef] [PubMed]
29. Pitt, J.J. Principles and applications of liquid chromatography-mass spectrometry in clinical biochemistry. *Clin. Biochem. Rev.* **2009**, *30*, 19–34. [PubMed]
30. Sawa, T.; Tatemichi, M.; Akaike, T.; Barbin, A.; Ohshima, H. Analysis of urinary 8-nitroguanine, a marker of nitrative nucleic acid damage, by high-performance liquid chromatography-electrochemical detection coupled with immunoaffinity purification: Association with cigarette smoking. *Free Radic. Biol. Med.* **2006**, *40*, 711–720. [CrossRef] [PubMed]
31. Lin, H.J.; Chen, S.T.; Wu, H.Y.; Hsu, H.C.; Chen, M.F.; Lee, Y.T.; Wu, K.Y.; Chien, K.L. Urinary biomarkers of oxidative and nitrosative stress and the risk for incident stroke: A nested case-control study from a community-based cohort. *Int. J. Cardiol.* **2015**, *183*, 214–220. [CrossRef] [PubMed]
32. Wang, P.W.; Chen, M.L.; Huang, L.W.; Yang, W.; Wu, K.Y.; Huang, Y.F. Nonylphenol exposure is associated with oxidative and nitrative stress in pregnant women. *Free Radic. Res.* **2015**, *49*, 1469–1478. [CrossRef] [PubMed]
33. Dizdaroglu, M. Facts about the artifacts in the measurement of oxidative DNA base damage by gas chromatography mass spectrometry. *Free Radic. Res.* **1998**, *29*, 551–563. [CrossRef] [PubMed]
34. Levrand, S.; Pesse, B.; Feihl, F.; Waeber, B.; Pacher, P.; Rolli, J.; Schaller, M.D.; Liaudet, L. Peroxynitrite is a potent inhibitor of NF-κB activation triggered by inflammatory stimuli in cardiac and endothelial cell lines. *J. Biol. Chem.* **2005**, *280*, 34878–34887. [CrossRef] [PubMed]
35. Nakae, D.; Mizumoto, Y.; Kobayashi, E.; Noguchi, O.; Konishi, Y. Improved genomic/nuclear DNA extraction for 8-hydroxydeoxyguanosine analysis of small amounts of rat liver tissue. *Cancer Lett.* **1995**, *97*, 233–239. [CrossRef]
36. Chao, M.R.; Wang, C.J.; Yen, C.C.; Yang, H.H.; Lu, Y.C.; Chang, L.W.; Hu, C.W. Simultaneous determination of N7-alkylguanines in DNA by isotope-dilution LC-tandem MS coupled with automated solid-phase extraction and its application to a small fish model. *Biochem. J.* **2007**, *402*, 483–490. [CrossRef] [PubMed]
37. Hu, C.W.; Chao, M.R.; Sie, C.H. Urinary analysis of 8-oxo-7,8-dihydroguanine and 8-oxo-7,8-dihydro-2'-deoxyguanosine by isotope-dilution LC-MS/MS with automated solid-phase extraction: Study of 8-oxo-7,8-dihydroguanine stability. *Free Radic. Biol. Med.* **2010**, *48*, 89–97. [CrossRef] [PubMed]

**Sample Availability:** Samples of 8-nitroguanine and [$^{13}C_2$,$^{15}N$]-8-nitroguanine are available from the authors.

*molecules*

MDPI

*Article*

# Efficient Separation of Four Antibacterial Diterpenes from the Roots of *Salvia Prattii* Using Non-Aqueous Hydrophilic Solid Phase Extraction Followed by Preparative High-Performance Liquid Chromatography

Jun Dang [1,2,†], Yulei Cui [1,2,†], Jinjin Pei [1,3], Huilan Yue [1,2], Zenggen Liu [1,2], Weidong Wang [1,2], Lijin Jiao [1,2], Lijuan Mei [1,2], Qilan Wang [1,2], Yanduo Tao [1,2,*] and Yun Shao [1,2,*]

[1] Key Laboratory of Tibetan Medicine Research, Northwest Institute of Plateau Biology, Chinese Academy of Sciences, Xining 810008, China; dangjun@nwipb.cas.cn (J.D.); m17701159965@163.com (Y.C.); jinjinpeislg@163.com (J.P.); hlyue@nwipb.cas.cn (H.Y.); lzg2005sk@126.com (Z.L.); wangweidong315@mails.ucas.ac.cn (W.W.); jiaolijin15@mails.ucas.ac.cn (L.J.); meilijuan111@163.com (L.M.); wql@nwipb.cas.cn (Q.W.)

[2] Qinghai Provincial Key Laboratory of Tibetan Medicine Research, Xining 810008, China

[3] Shaanxi Key Laboratory of Bio-Resources, Shaanxi University of Technology, Hanzhong 723000, China

* Correspondence: chemi_ttm_2012@163.com (Y.T.); shaoyun11@126.com (Y.S.); Fax: +86-971-6143282 (Y.T.)

† These authors contributed equally to this work.

Received: 15 January 2018; Accepted: 9 March 2018; Published: 9 March 2018

**Abstract:** An efficient preparative procedure for the separation of four antibacterial diterpenes from a *Salvia prattii* crude diterpenes-rich sample was developed. Firstly, the XION hydrophilic stationary phase was chosen to separate the antibacterial crude diterpenes-rich sample (18.0 g) into three fractions with a recovery of 46.1%. Then, the antibacterial fractions I (200 mg), II (200 mg), and III (150 g) were separated by the Megress C18 preparative column, and compounds tanshinone IIA (80.0 mg), salvinolone (62.0 mg), cryptotanshinone (70.0 mg), and ferruginol (68.0 mg) were produced with purities greater than 98%. The procedure achieved large-scale preparation of the four diterpenes with high purity, and it could act as a reference for the efficient preparation of active diterpenes from other plant extracts.

**Keywords:** *Salvia prattii*; antibacterial diterpenes; hydrophilic solid-phase extraction; preparative high-performance liquid chromatography

## 1. Introduction

*Salvia prattii* (*S. prattii*), acknowledged as an alternative for *Salvia miltiorrhiza*, is extensively utilized in traditional Tibetan medicine. Previous chemical investigations have proved that *Salvia* species possess two main classes of biologically active substances: phenylpropanoids and diterpenes [1–3]. Diterpenes, the principal active constituents of other *Salvia* plants, have numerous pharmacological functions, including antibacterial [4], anti-inflammatory [5], and anticancer activities [6,7]. In our preliminary experiment, the crude diterpenes-rich sample of *S. prattii* displayed considerable antibacterial activity against *Staphylococcus aureus* (MIC: 125 µg/mL), *Pseudomonas aeruginosa* (MIC: 125 µg/mL), and *Acinetobacter baumannii* (MIC: 250 µg/mL). To identify the main antibacterial constituents of the sample, it is desirable to obtain the diterpenes in adequate purity and quantity. Thus, the objective of this work is to develop an efficient process for the purification of diterpenes from the diterpenes-rich sample of *S. prattii*.

To date, the separation of diterpenes from *Salvia* plants depends on gel and silica gel open column chromatography [8,9]. However, such methods have numerous drawbacks, such as low yields, being time-consuming, producing a large quantity of solvent waste, and non-suitability for large-scale industrial production. In recent times, high-speed counter-current chromatography, a liquid-liquid chromatographic method, was proposed for the isolation of diterpenes from *Salvia* plants [10–12]. Even though this technique offers high-separation efficiency, it requires several hours, rather than minutes needed for preparative high-performance liquid chromatography (prep-HPLC).

Prep-HPLC, is considered as an efficient technique for the separation and purification of phenols, coumarins, flavonoids, and glycosides from intricate mixtures like traditional Tibetan medicines [13–15]. It is preferred over other chromatography methods, owing to higher efficiency, greater resolution, and better reproducibility through online monitoring and automatic control [16–18]. Consequently, prep-HPLC has been drawing ever-increasing attention from phytochemists and the pharma industry. However, a crude extract cannot be directly subjected to prep-HPLC separation; other separation techniques are usually required for enrichment of the main compounds of the crude extract. Hydrophilic interaction liquid chromatography solid-phase extraction (HILIC-SPE), which employs stationary phases with polar functional groups bonded to silica gel surface, has been extensively used to enrich compounds of interest from natural products due to its applicability, ease of use and regeneration, as well as complementary selectivity to reversed-phase liquid chromatography [19]. A few studies reported the separation of diterpenes from *S. miltiorrhiza* using high-speed counter-current chromatography [20–22], but no reports mentioned the separation of diterpenes by a combination of HILIC-SPE and prep-HPLC. Hence, this study aimed to develop a valuable protocol for the purification of four antibacterial diterpenes from a diterpenes-rich *S. prattii* crude sample. The developed protocol succeeded in achieving large-scale preparation of four highly pure diterpenes from the crude sample of *S. prattii*, paving way for the potential development of antibacterial drugs.

## 2. Experimental

### 2.1. Apparatus

The prep-HPLC experiment was performed on a Hanbon DAC-50 prep-HPLC system (Hanbon Science & Technology Co., Ltd., Huai'an, China). The system consisted of a DAC-50 Megress C18 dynamic axial compression column, two prep-HPLC NP7000 pumps, a sample loop of 20.0 mL, a DM-A Dynamic Mixer, a NU3000 UV/Vis detector and an EasyChrom workstation.

The HPLC analysis was carried out on an Agilent 1200 instrument (Agilent Technologies Co., Ltd., Santa Clara, CA, USA) consisting of a G1311A pump, a G1315D UV/Vis detector, a G1316A thermostat, an autosampler and an Agilent workstation. ESI-MS spectra were recorded on an API 2000 mass spectrometer (AB SCIEX, Milwaukee, WI, USA). The NMR spectra were recorded on Bruker Avance 600 MHz (Bruker, Karlsruhe, Germany) spectrometer using tetramethylsilane (TMS) as the internal standard.

### 2.2. Reagents and Stationary Phases

Analytical grade 95% ethanol, *n*-hexane, ethanol, and methanol utilized for the sample extraction, as well as HILIC-SPE and prep-HPLC were ordered from the Tianjin Chemical Factory (Tianjin, China). Chromatographic grade *n*-hexane, ethanol, methanol and acetonitrile employed for the HPLC analysis were bought from Concord Chemical Ltd. (Tianjin, China). Water was purified through a PAT-125 laboratory ultrapure water system from Chengdu ultra Tech (Chengdu, China).

The XION (40–60 μm) and Megress C18 (10 μm) stationary phases were purchased from Acchrom Technologies Co., Ltd. (Beijing, China) and Hanbon Science & Technology Co., Ltd. (Huai'an, China), respectively. The XION (250 mm × 4.6 mm, 40–60 μm) and Megress C18 (250 mm × 4.6 mm, 10 μm) analytical columns were obtained from Acchrom (Beijing, China) and Hanbon Science & Technology

Co., Ltd. (Huai'an, China), respectively. Silica gel (250 mm × 4.6 mm, 40–60 μm) and XAqua C3 (250 mm × 4.6 mm, 5 μm) analytical columns were obtained from Acchrom (Beijing, China).

### 2.3. Preparation of the Crude Sample

The roots of *S. prattii* have been obtained from Yushu in Qinghai province, China (September 2016) and authenticated by Prof. Li-Juan Mei of the Northwest Institute of Plateau Biology, Chinese Academy of Sciences. A sample (NWIPB-SPH-2016-11-14) was handed over to Qinghai-Tibetan Plateau Museum of Biology (QPMB). The dried and milled samples (1.2 kg) were extracted thrice for 2 days using 95% ethanol (12.0 L for each extraction) at room temperature. The extracts were combined (36.0 L) and concentrated at 60 °C using a rotary evaporator. The partially dried concentrate (approximately 0.5 L) was suspended in distilled water (2.0 L); the suspension was subsequently loaded onto a preprocessed middle chromatogram isolated gel (MCI) column (10 cm × 100 cm, 2 kg), washed with 40% ethanol (12 L), eluted with of 80% ethanol (12 L) and further dried to yield 18.0 g of crude diterpenes-rich sample for ensuing HILIC-SPE pre-separation.

### 2.4. HILIC-SPE Pre-Separation

The crude diterpenes-rich sample was dissolved in methanol, mixed with polyamide and dried using a rotary evaporator. Afterwards, the solid mixture was loaded onto a XION solid-phase extraction medium-pressure column (300 mm × 50 mm, containing 297.7 g solid-phase XION) and eluted with four column volumes of *n*-hexane/ethanol (20:0, 19:1, 18:2, 16:4 and 14:6 *v/v*), successively. The eluent from the HILIC-SPE column was collected in 100 mL fractions and analyzed by HPLC using a XION (250 mm × 4.6 mm, 40–60 μm) analytical column. The eluents with the same composition were collected and combined according to the HPLC analysis. Finally, the fractions eluted with *n*-hexane/ethanol 16:4 and 14:6 *v/v* gave fraction I (2.8 g), fraction II (3.4 g), and fraction III (2.1 g), respectively. The three fractions were stored in a refrigerator for subsequent preparative separation.

### 2.5. Antibacterial Activity

*Staphylococcus aureus* (ATCC 25923), *Pseudomonas aeruginosa* (ATCC 27853), and *Acinetobacter baumannii* (obtained from the People's Liberation Army (PLA) General Hospital) were used as the instruction strains for the antibacterial activity assay. Mueller-Hinton broth was used to culture bacteria and an increase in optical density at 600 nm was used to monitor growth. The two-fold serial dilutions of the active extracts and diterpenes (dissolved in dimethyl sulfoxide, DMSO) were added into the sensitive strains, respectively. The minimum inhibitory concentration (MIC), defined as the lowest concentration of the active extracts and diterpenes needed to inhibit the growth of the sensitive strains, was observed following incubation at 30 °C for 18 h according to the Clinical and Laboratory Standards Institute (CLSI, Wayne, PA, USA, 2008). The mid-exponential broth of sensitive strains treated without the extracts, diterpenes and DMSO were considered as the negative control. The growth of only DMSO-treated sensitive strains was monitored to eliminate the effect of DMSO. The mid-exponential broth of sensitive strains treated with the antibiotic cefotaxime and vancomycin (1.0 mg/mL) were used as positive controls.

### 2.6. Purification of the Main Diterpenes by Prep-HPLC

The purification of diterpenes was performed on a Hanbon DAC-50 prep-HPLC system. Fractions I, II and III were dissolved in methanol and injected onto a DAC-50 dynamic axial compression column containing the Megress C18 stationary phase (flow rate: 60 mL/min; injection volume: 5.0 mL). The mobile phases consisted of 0.2% *v/v* formic acid in water and 0.2% *v/v* formic acid in methanol at different ratios (15:85 for fraction I, 20:80 for fractions II and III). The effluent was analyzed using a UV/Vis detector at 254 nm and was manually obtained based on the chromatograms. The collected fractions were subsequently evaporated to dryness in reduced pressure at 60 °C.

*2.7. HPLC Analysis and Identification of the Separated Diterpenes*

HPLC analysis of the separated diterpenes was carried out at 25 °C, on a XAqua C3-column (flow rate: 1.0 mL/min) and the chromatogram was recorded at 254 nm. Water and methanol were the mobile phases A and B, respectively. The gradient elution steps were as follows: 0–30 min, 75–85% B.

The chemical structures of the separated diterpenes were established by UV, Mass, $^1$H-NMR and $^{13}$C-NMR spectrometry. The UV spectra were recorded using the DAD detector of the Agilent 1200 system. ESI-MS spectra were recorded on an API 2000 mass spectrometer in positive ion mode, whereas NMR spectra were recorded on a Bruker Avance III 600 MHz spectrometer.

## 3. Results and Discussion

*3.1. HILIC-SPE Column Chromatography Fractionation and Antibacterial Activity Screening*

To simplify the development of the reversed-phase prep-HPLC method and to improve the life-span of the reversed-phase stationary phase, the crude diterpenes-rich extract with complex composition usually requires pretreatment. To select an appropriate pretreatment, two chromatographic stationary phases i.e., the bare silica gel and the XION stationary phases, in three separation modes were tested for the separation of the crude extract; the representative separation chromatograms are shown in Figure 1. As observed in Figure 1A,B, the main diterpenes had inferior resolution on the bare silica gel stationary phase compared to that on the XION stationary phase under the same elution conditions (*n*-hexane/ethanol solvent system). The main diterpenes showed weak retention on the XION stationary phase with the mobile phases of 0.2% *v*/*v* formic acid in acetonitrile and in water (Figure 1C). According to the manufacturer, XION is a cysteine-bond silica gel stationary phase, and cysteine is a polar group, which gives the XION stationary phase the retention behavior of normal-phase chromatography and hydrophilic interaction chromatography [23,24]. Thus, the XION stationary phase should be employed for sample pretreatment under the normal-phase mode due to the favorable separation profile (Figure 1B).

An analytical column (250 mm × 4.6 mm, 40–60 μm) for hydrophilic interaction chromatography is usually packed with 2.1 g of stationary phase ($\rho$ was approximately 0.5 g/mm$^3$ under the conditions of high pressure), and one column volume is 2.1 mL (one column volume: weight of the stationary phase = 1 mL:1 g). The calculations used the following equation:

$$\frac{\rho_A \pi R_A^2 H_A}{\rho_A \pi R_P^2 H_P} = \frac{m_A}{m_P} \tag{1}$$

where $\rho_A$ and $\rho_P$ are stationary phase packing densities of the analytical column and HILIC-SPE column (under the conditions of high pressure, $\rho_A = \rho_P$), respectively; $R_A$ and $R_P$ are the diameters of the analytical column (4.6 mm) and HILIC-SPE column (50 mm), respectively; $H_A$ and $H_P$ are the column lengths of the analytical column (250 mm) and HILIC-SPE column (actual packing length was 300 mm), respectively; similarly, $m_A$ and $m_P$ are stationary phase weights of the analytical column and the HILIC-SPE column, respectively. For the same stationary phase, the packing density in the analytical column and HILIC-SPE column was uniform under the conditions of high pressure. Thus, the above equation could be simplified as:

$$\frac{R_A^2 H_A}{R_P^2 H_P} = \frac{m_A}{m_P} \tag{2}$$

The calculations showed that the HILIC-SPE column (300 mm × 50 mm, 40–60 μm) should be packed with 297.7 g of stationary phase, and one column volume is 297.7 mL. Therefore, the sample loaded onto the XION solid-phase extraction column was eluted with 1190.8 mL (four column volumes) of *n*-hexane/ethanol (20:0, 19:1, 18:2, 16:4, and 14:6 *v*/*v*, successively); the same gradient elution was used for the XION analytical column and HILIC-SPE column. The HPLC analysis results revealed that the diterpene fractions were mainly present in the eluates of the 16:4 and 14:6 *v*/*v* *n*-hexane/ethanol.

Figure 2 shows the target fractions I (Figure 2A), II (Figure 2B), and III (Figure 2C). By comparing the Figures 1 and 2, it could be seen that the diterpenes were divided into three groups (fractions I, II and III) with the *n*-hexane/ethanol mobile phase on the XION stationary phase. Following the HILIC-SPE column chromatography, 2.8 g of fraction I, 3.4 g of fraction II, and 2.1 g of fraction III were obtained from 18.0 g of the antibacterial crude diterpenes-rich sample.

The antibacterial activities (MIC in μg/mL) of the tested fractions I–III are shown in Table 1. Fractions I–III displayed higher antimicrobial activity compared to the crude diterpenes-rich sample against *Staphylococcus aureus*, *Pseudomonas aeruginosa*, and *Acinetobacter baumannii*. Therefore, it is of interest to identify the antibacterial compounds from the bioactive fractions I, II, and III through further separation.

**Table 1.** Antimicrobial activity of the crude diterpene-rich sample, fractions and, compounds (MIC in μg/mL).

| Bacteria | | | | | Cf |
|---|---|---|---|---|---|
| Crude diterpene-rich sample | | | | | |
| *Staphylococcus aureus* | 125 | | | | 0.5 |
| *Pseudomonas aeruginosa* | 125 | | | | 7.5 |
| *Acinetobacter baumannii* | 250 | | | | 12.5 |
| Fractions I, II and III | | | | | |
| | I | II | III | | |
| *Staphylococcus aureus* | 100 | 125 | 50 | | 0.5 |
| *Pseudomonas aeruginosa* | 100 | 150 | 50 | | 7.5 |
| *Acinetobacter baumannii* | 150 | 150 | 50 | | 12.5 |
| Compounds | | | | | |
| | 1 | 2 | 3 | 4 | |
| *Staphylococcus aureus* | 25 | 30 | 25 | 10 | 0.5 |
| *Pseudomonas aeruginosa* | 20 | 50 | 30 | 15 | 7.5 |
| *Acinetobacter baumannii* | 25 | 50 | 25 | 15 | 12.5 |

(1) tanshinone IIA; (2) salvinolone; (3) cryptotanshinone; (4) ferruginol. Cf: Vancomycin for *Staphylococcus aureus*; cefotaxime for *Pseudomonas aeruginosa* and *Acinetobacter baumannii*.

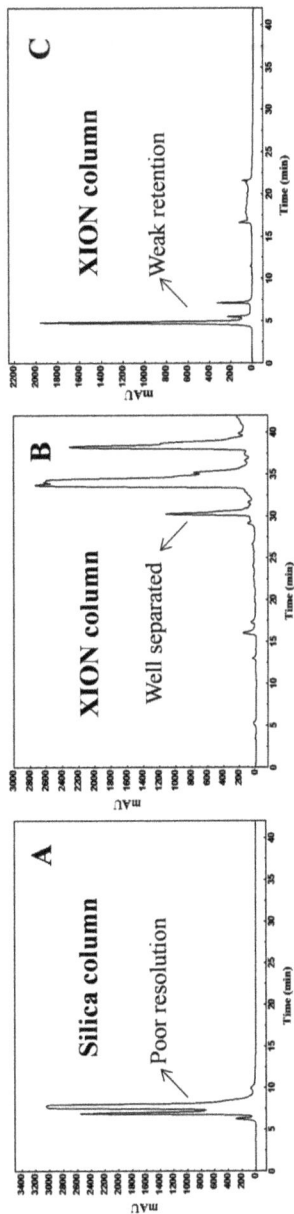

**Figure 1.** Representative analytical chromatograms of *S. prattii* crude extract on a silica analytical column (**A**) and XION analytical column (**B**,**C**). The conditions of (**A**,**B**) were the same: mobile phase A: *n*-hexane, and mobile phase B: ethanol; gradient: 0–8.4 min, 0% B; 8.4–16.8 min, 5%~5% B; 16.8–25.2 min, 10%~10% B; 25.2–33.6 min, 20%~20% B; 33.6–42 min, 30%~30% B. Conditions of (**C**): mobile phase A: 0.2% *v*/*v* formic acid in acetonitrile, and mobile phase B: 0.2% *v*/*v* formic acid in water; gradient: 0–10 min, 0%~0% B; 10–56 min, 0%~95% B. The other conditions were the same: monitoring wavelength: 254 nm; flow rate: 1 mL/min; injection volume: 10 μL; column temperature: 25 °C.

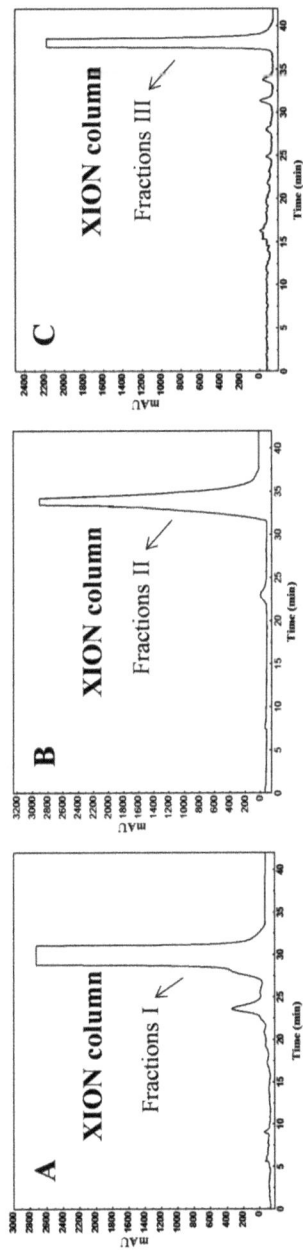

**Figure 2.** The analytical chromatograms of target fractions I (**A**), II (**B**) and III (**C**) on a XION analytical column. Conditions of (**A**), (**B**), and (**C**): mobile phase A: *n*-hexane, and mobile phase B: ethanol; gradient: 0–8.4 min, 0%~0% B; 8.4–16.8 min, 5%~5% B; 16.8–25.2 min, 10%~10% B; 25.2–33.6 min, 20%~20% B; 33.6–42 min, 30%~30% B. Other conditions: monitoring wavelength: 254 nm; flow rate: 1 mL/min; injection volume: 10 μL; column temperature: 25 °C.

*3.2. Purification of Diterpenes by Reversed-Phase Preparative High-Performance Liquid Chromatography*

To attain a good separation profile, the stationary phases used in the analytical HPLC and preparative liquid chromatography were the same. The elution conditions of fractions I, II, and III were standardized on the Megress C18 analytical column, whereas the linear magnifying methodology has been utilized to transform the analytical flow rate to the preparative level using the following equation:

$$\frac{R_A^2}{R_P^2} = \frac{F_A}{F_P} \qquad (3)$$

where $R_A$ and $R_P$ are the diameters of the analytical column (4.6 mm) and preparative column (50 mm), respectively. Similarly, $F_A$ and $F_P$ are the flow rates of the analytical column and preparative column, respectively. For convenience, the isocratic elution conditions of same sample solutions were optimized on the Megress C18 analytical column (250 mm × 4.6 mm, 10 μm) at a flow rate of 0.5 mL/min ($F_A$). The comparable flow rate in the Megress C18 preparative column (250 mm × 50 mm, 10 μm) has been determined to be 59.1 mL/min ($F_P$), hence, 60.0 mL/min was used for convenience. For the loading amount, a similar linear magnifying technology was employed with the equation:

$$\frac{R_A^2}{R_P^2} = \frac{M_A}{M_P} \qquad (4)$$

For the same concentration (40.0 mg/mL for fractions I and II; 30.0 mg/mL for fraction III) of the sample solutions on the analytical and preparative columns, the equation could be simplified as:

$$\frac{R_A^2}{R_P^2} = \frac{V_A}{V_P} \qquad (5)$$

where $M_A$ and $M_P$ are the loading amounts of the analytical and preparative columns, respectively. Similarly, $V_A$ and $V_P$ are the injection volumes of the analytical and preparative columns, respectively. The injection volumes (loading amount) of the analytical column and the preparative columns were 10 μL ($V_A$) and 1181.5 μL (1.181 mL), respectively. However, this volume is too low to meet the requirements for the highly proficient preparation, and the preparative injection volume was larger but smaller than that of the sample loop (5.0 mL for fractions I, II, and III in actual operation).

Under the optimized conditions, the separation chromatograms of fractions I–III on the analytical level and preparative level were shown in Figure 3. The optimized analytical (Figure 3A–C) and preparative (Figure 3D–F) chromatographic profiles were similar, except for a slight increase in the retention times of peaks 1–4 on the prep-chromatograms. For peaks 2 and 3 (Figure 3B,E), the resolution on the analytical and preparative chromatograms were $R_{SA}$ = 2($R_{t2}$ − $R_{t1}$)/($W_{t2}$ + $W_{t1}$) = 2.55 and $R_{SP}$ = 1.2, respectively, due to sample-overloading and the preparative column diffusion effect. The target compound chromatographic peaks were collected based on the UV absorption intensities considerably enhance the purity. Overall, 200.0 mg each of fractions I and II as well as 150.0 mg of fraction III were separated to obtain 80.0 mg of the peak 1 fraction, 62.0 mg of the peak 2 fraction, 70.0 mg of the peak 3 fraction, and 68.0 mg of the peak 4 fraction with recoveries of 40.0%, 66.0%, and 45.3%, respectively.

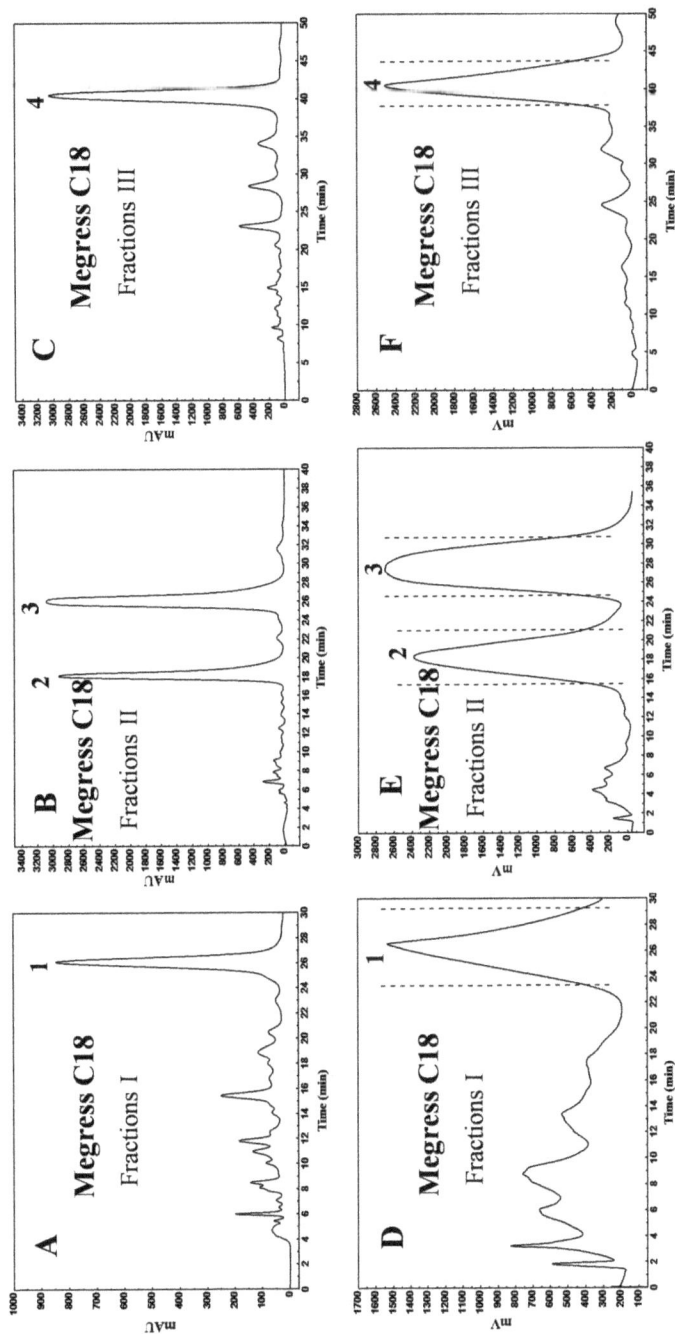

**Figure 3.** Analytical chromatograms of target fractions I (**A**), II (**B**), and III (**C**) on a Megress C18 analytical column and preparative chromatograms of target fractions I (**D**), II (**E**), and III (**F**) on a Megress C18 preparative column. The mobile phases for fractions I (**A**,**D**), II (**B**,**E**), and III (**C**,**F**) were the same: A: 0.2% *v*/*v* formic acid in water, and B: 0.2% *v*/*v* formic acid in methanol. Isocratic: 0–30 min, 85%~85% B for fraction I (**A**,**D**). Isocratic: 0–40 min, 80%~80% B for fraction II (**B**,**E**). Isocratic: 0–50 min, 80%~80% B for fraction III (**C**,**F**). The other conditions were the same: monitoring wavelength: 254 nm; flow rate: 1 mL/min; injection volume: 10 μL; column temperature: room temperature.

The purities of the four isolated diterpenes were determined by HPLC on XAqua C3 analytical column (250 mm × 4.6 mm, 5 μm). The results are given in Figure 4: Figure 4A–D show that the compounds possess purity greater than 98% (98.6% for compound 1, 99.9% for compound 2, 99.1% for compound 3, and 99.7% for compound 4). The UV spectra of the four diterpenes are included in the insets of Figure 4 (Figure 4A$_1$–D$_1$); they were in accordance with those in previous publications [23–26]. In addition, the antibacterial activity of the four isolated diterpenes was also tested (Table 1): Compound 4 showed the strongest activity (MIC: 10–15 μg/mL) against *Staphylococcus aureus*, *Pseudomonas aeruginosa*, and *Acinetobacter baumannii*; whereas compounds 1–3 also displayed potent activity against the same species (MIC: 20–50 μg/mL).

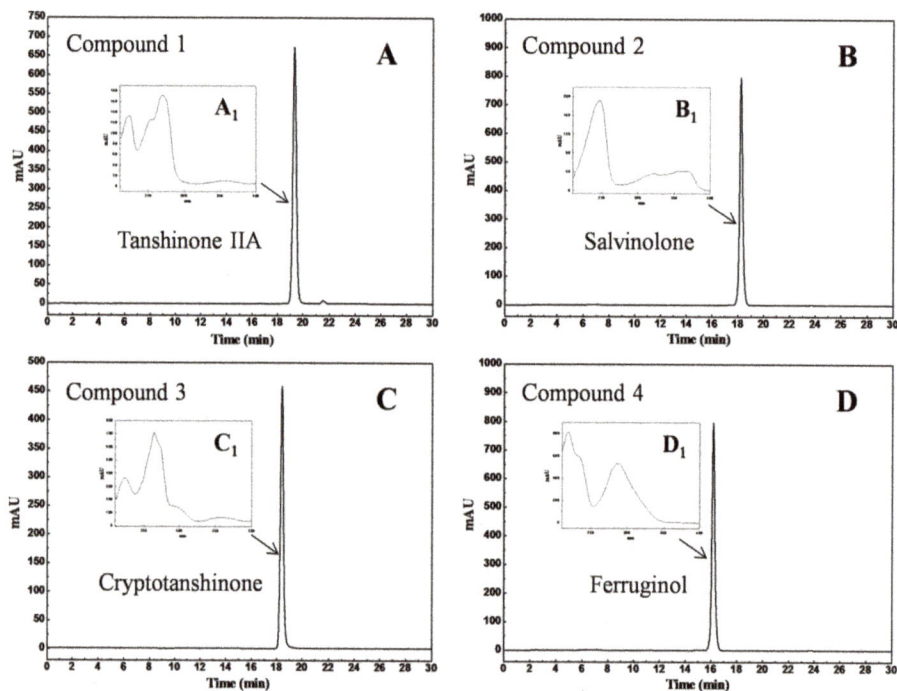

**Figure 4.** Purity analysis chromatograms (**A–D**) of the isolated compounds on an XAqua C3 analytical column and their UV spectra (**A$_1$–D$_1$**). Conditions: mobile phase A: water, and mobile phase B: methanol; gradient: 0–30 min, 75%~85% B. Other conditions: monitoring wavelength: 254 nm; flow rate: 1.0 mL/min; injection volume: 10 μL; column temperature: 25 °C.

The chemical structures of the isolated diterpenes have been determined by ESI-MS, $^1$H-NMR, and $^{13}$C-NMR spectroscopy. By comparing the spectra of the compounds with literature data, we concluded that compounds **1–4** represent tanshinone IIA, salvinolone, cryptotanshinone, and ferruginol. The chemical structures of tanshinone IIA (compound **1**), salvinolone (**2**), cryptotanshinone (**3**), and ferruginol (**4**) are shown in Figure 5 and the detailed data of the compounds are given below.

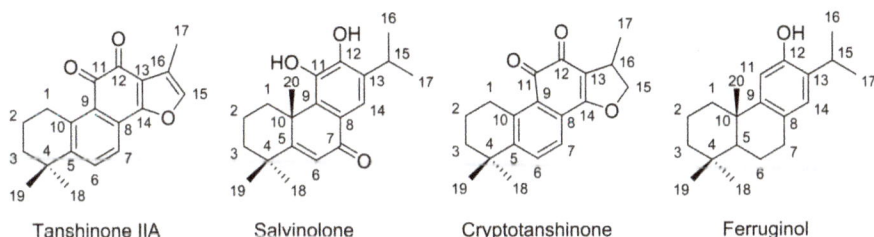

**Figure 5.** The chemical structures of tanshinone IIA (compound **1**), salvinolone (**2**), cryptotanshinone (**3**), and ferruginol (**4**).

Compound **1**: Red powder, ESI-MS *m/z*: 295.3 [M + H]⁺. ¹H-NMR (600 MHz, CDCl₃): δ 7.63 (1H, d, *J* = 8.1 Hz, H-6), 7.55 (1H, d, *J* = 8.1 Hz, H-7), 7.22 (1H, s, H-16), 3.18 (2H, t, *J* = 6.4 Hz, H-1), 2.26 (3H, s, H-17), 1.79 (2H, m, H-2), 1.66 (2H, m, H-3), 1.31 (6H, s, H-18, 19); ¹³C-NMR (151 MHz, CDCl₃): 183.8 (C-11), 175.9 (C-12), 161.9 (C-14), 150.3 (C-10), 144.6 (C-5), 141.4 (C-15), 133.6 (C-6), 127.6 (C-8), 126.6 (C-9), 121.3 (C-16), 120.4 (C-7), 120.0 (C-13), 38.0 (C-3), 34.8 (C-4), 32.0 (C-18), 32.0 (C-19), 30.0 (C-1), 19.3 (C-2), 9.0 (C-17). The ESI-MS, ¹H-NMR and ¹³C-NMR data for compound **1** agree with the literature data reported for tanshinone IIA [25].

Compound **2**: Red powder, ESI-MS *m/z*: 315.3 [M + H]⁺. ¹H-NMR (600 MHz, CDCl₃): δ 7.72 (1H, s, H-14), 6.45 (1H, s, H-6), 3.28 (2H, m, H-15), 1.67 (3H, s, H-20), 1.36 (3H, s, H-18), 1.32 (3H, d, *J* = 6.5 Hz, H-17), 1.30 (3H, d, *J* = 6.5 Hz, H-16), 1.26 (3H, s, H-19); ¹³C-NMR (151 MHz, CDCl₃): 185.8 (C-7), 175.2 (C-5), 144.6 (C-12), 141.5 (C-11), 137.5 (C-9), 132.6 (C-13), 124.4 (C-6), 123.7 (C-8), 115.8 (C-14), 42.3 (C-10), 40.5 (C-3), 38.3 (C-4), 34.5 (C-1), 33.3 (C-20), 29.6 (C-19), 27.6 (C-18), 25.1 (C-15), 22.8 (C-16), 22.6 (C-17), 18.9 (C-2). The ESI-MS, ¹H-NMR and ¹³C-NMR data for compound **2** agree with the literature data reported for salvinolone [26].

Compound **3**: Orange powder, ESI-MS *m/z*: 297.4 [M + H]⁺. ¹H-NMR (600 MHz, CDCl₃): δ 7.63 (1H, d, *J* = 8.1 Hz, H-6), 7.49 (1H, d, *J* = 8.1 Hz, H-7), 4.88 (1H, t, *J* = 9.4 Hz, H-15), 4.36 (1H, dd, *J* = 9.2 and 6.0 Hz, H-16), 3.21 (2H, t, *J* = 6.4 Hz, H-1), 1.78 (2H, m, H-2), 1.65 (2H, m, H-3), 1.35 (3H, d, *J* = 6.8 Hz, H-17), 1.30 (6H, s, H-18, 19); ¹³C-NMR (151 MHz, CDCl₃): 184.4 (C-11), 175.8 (C-12), 171.0 (C-14), 152.5 (C-10), 143.8 (C-5), 132.7 (C-6), 128.5 (C-8), 126.3 (C-9), 122.6 (C-7), 118.4 (C-13), 81.6 (C-15), 37.9 (C-3), 35.0 (C-4), 34.7 (C-16), 32.0 (C-18), 32.0 (C-19), 29.8 (C-1), 19.2 (C-2), 19.0 (C-17). The ESI-MS, ¹H-NMR and ¹³C-NMR data for compound **3** agree with the literature data reported for cryptotanshinone [27].

Compound **4**: Yellow powder, ESI-MS *m/z*: 287.4 [M + H]⁺. ¹H-NMR (600 MHz, CDCl₃): δ 6.84 (1H, s, H-14), 6.63 (1H, s, H-10), 3.13 (1H, m, H-15), 2.87 (1H, dd, *J* = 16.6 and 6.6 Hz, H-7a), 2.77 (1H, m, H-7b), 1.25 (3H, d, *J* = 7.4 Hz, H-17), 1.23 (3H, d, *J* = 7.2 Hz, H-16), 1.18 (1H, s, H-20), 0.95 (3H, s, H-19), 0.93 (3H, s, H-18); ¹³C-NMR (151 MHz, CDCl₃): 150.8 (C-12), 148.8 (C-9), 131.5 (C-13), 127.4 (C-8), 126.7 (C-14), 111.1 (C-11), 50.5 (C-5), 41.8 (C-3), 39.0 (C-1), 37.6 (C-10), 33.6 (C-4), 33.5 (C-18), 29.9 (C-7), 26.9 (C-15), 24.9 (C-20), 22.9 (C-17), 22.7 (C-16), 21.8 (C-19), 19.5 (C-2), 19.4 (C-6). The ESI-MS, ¹H-NMR and ¹³C-NMR data for compound **4** agree with the literature data reported for ferruginol [28].

## 4. Concluding Remarks

An efficient preparative procedure involving non-aqueous HILIC-SPE followed by prep-HPLC was developed for the preparation of antibacterial tanshinone IIA, salvinolone, cryptotanshinone, and ferruginol from a crude diterpenes-rich sample of the roots of *S. prattii*. Initially, HILIC-SPE was used to separate 18.0 g of crude diterpenes-rich sample into three fractions (2.8 g of fraction I, 3.4 g of fraction II, and 2.1 g of fraction III). Then, a DAC-50 prep-HPLC column containing Megress C18 stationary phase was used to isolate the diterpenes from the three fractions. The separation of each fraction on the Megress C18 was performed by optimizing the separation conditions on the

Megress C18 analytical column and transforming the conditions to a Megress C18 preparative column. The coeluted diterpenes from the HILIC-SPE stationary phase were separated at large-scale sample amounts on the Megress C18 reversed stationary phase, brought about by the diverse sample separation mechanisms of both the stationary phases. Owing to the different selectivities and the optimized collection mode, 80.0 mg of tanshinone IIA, 62.0 mg of salvinolone, 70.0 mg of cryptotanshinone, and 68.0 mg of ferruginol at greater than 98% purity were prepared from 200.0 mg each of fractions I and II as well as 150.0 mg of fraction III with a single preparation. The purified diterpenes were identified as tanshinone IIA, salvinolone, cryptotanshinone, and ferruginol by UV, Mass, and NMR spectroscopy. Therefore, the present work offers an excellent protocol for the preparation of active diterpenes from plant sources.

**Acknowledgments:** The authors gratefully acknowledge the financial support by the High and New Technology Research and Development Planning, Significant Science & Technological Project of Qinghai Province (2014-GX-A3A), the Natural Science Funds of Qinghai Province (2016-ZJ-933Q), the Project of Discovery, Evaluation and Transformation of Active Natural Compounds, Strategic Biological Resources Service Network Programme of Chinese Academy of Sciences (ZSTH-027), the West Light Foundation of the Chinese Academy of Sciences for Doctors (2014) and the Youth Innovation Promotion Association of Chinese Academy of Sciences (2017471).

**Author Contributions:** Yanduo Tao and Yun Shao conceived and designed the research framework; Lijuan Mei and Zenggen Liu prepared the plant sample; Jun Dang, Yulei Cui, Jinjin Pei, Huilan Yue, Weidong Wang, Lijin Jiao, Qilan Wang performed the experiments; Jun Dang wrote the paper; Huilan Yue and Zenggen Liu made revisions to the final manuscript. All authors have read and approved the final manuscript.

**Conflicts of Interest:** The authors declare no conflict of interest.

# References

1. Al-Qudah, M.A.; Al-Jaber, H.I.; Abu Zarga, M.H.; Abu Orabi, S.T. Flavonoid and phenolic compounds from *Salvia palaestina* L. growing wild in Jordan and their antioxidant activities. *Phytochemistry* **2014**, 99, 115–120. [CrossRef] [PubMed]

2. Wang, Z.B.; Cao, B.C.; Yu, A.M.; Zhang, H.Q.; Qiu, F.P. Ultrasound-assisted ionic liquid-based homogeneous liquid-liquid microextraction high-performance liquid chromatography for determination of tanshinones in *Salvia miltiorrhiza* Bge. root. *J. Pharm. Biomed. Anal.* **2015**, 104, 97–104. [CrossRef] [PubMed]

3. Dang, J.; Shao, Y.; Zhao, J.Q.; Mei, L.J.; Tao, Y.D.; Wang, Q.L.; Zhang, L. Two-dimensional hydrophilic interaction chromatography × reversed-phase liquid chromatography for the preparative isolation of potential anti-hepatitis phenylpropanoids from *Salvia prattii*. *J. Sep. Sci.* **2016**, 39, 3327–3332. [CrossRef] [PubMed]

4. Yang, Z.X.; Kitano, Y.; Chiba, K.; Shibata, N.; Kurokawa, H.; Doi, Y.; Arakawa, Y.; Tada, M. Synthesis of variously oxidized abietane diterpenes and their antibacterial activities against MRSA and VRE. *Bioorg. Med. Chem.* **2001**, 9, 347–356. [CrossRef]

5. Theoduloz, C.; Delporte, C.; Valenzuela-Barra, G.; Silva, X.; Cadiz, S.; Bustamante, F.; Pertino, M.W.; Schmeda-Hirschmann, G. Topical anti-inflammatory activity of new hybrid molecules of terpenes and synthetic drugs. *Molecules* **2015**, 20, 11219–11235. [CrossRef] [PubMed]

6. Chen, W.X.; Liu, L.; Luo, Y.; Odaka, Y.; Awate, S.; Zhou, H.Y.; Shen, T.; Zheng, S.Z.; Lu, Y.; Huang, S.L. Cryptotanshinone activates p38/JNK and inhibits Erk1/2 leading to caspase-independent cell death in tumor cells. *Cancer Prev. Res.* **2012**, 5, 778–787. [CrossRef] [PubMed]

7. Cheng, C.Y.; Su, C.C. Tanshinone IIA may inhibit the growth of small cell lung cancer H146 cells by up-regulating the Bax/Bcl-2 ratio and decreasing mitochondrial membrane potential. *Mol. Med. Rep.* **2010**, 3, 645–650. [PubMed]

8. Yang, J.; Choi, L.L.; Li, D.Q.; Yang, F.Q.; Zeng, L.J.; Zhao, J.; Li, S.P. Simultaneous analysis of hydrophilic and lipophilic compounds in *Salvia miltiorrhiza* by double-development HPTLC and scanning densitometry. *JPC J. Planar Chromat.* **2010**, 24, 257–263. [CrossRef]

9. Songsri, S.; Nuntawong, N. Cytotoxic labdane diterpenes from *Hedychium ellipticum* Buch. -Ham. ex Sm. *Molecules* **2016**, 21, 749. [CrossRef] [PubMed]

10. Meng, J.; Yang, Z.; Liang, J.L.; Zhou, H.; Wu, S.H. Comprehensive multi-channel multi-dimensional counter-current chromatography for separation of tanshinones from *Salvia miltiorrhiza* Bunge. *J. Chromatogr. A* **2014**, 1323, 73–81. [CrossRef] [PubMed]

11. Meng, J.; Yang, Z.; Liang, J.L.; Guo, M.Z.; Wu, S.H. Multi-channel recycling counter-current chromatography for natural product isolation: Tanshinones as examples. *J. Chromatogr. A* **2014**, *1327*, 27–38. [CrossRef] [PubMed]

12. Sun, A.L.; Zhang, Y.Q.; Li, A.F.; Meng, Z.L.; Liu, R.M. Extraction and preparative purification of tanshinones from *Salvia miltiorrhiza* Bunge by high-speed counter-current chromatography. *J. Chromatogr. B* **2011**, *879*, 1899–1904. [CrossRef] [PubMed]

13. Dang, J.; Tao, Y.D.; Shao, Y.; Mei, L.J.; Zhang, L.; Wang, Q.L. Antioxidative extracts and phenols isolated from Qinghai-Tibet Plateau medicinal plant *Saxifraga tangutica* Engl. *Ind. Crops Prod.* **2015**, *78*, 13–18. [CrossRef]

14. Cheng, G.J.S.; Li, G.K.; Xiao, X.H. Microwave-assisted extraction coupled with counter-current chromatography and preparative liquid chromatography for the preparation of six furocoumarins from *Angelica pubescentis* Radix. *Sep. Purif. Technol.* **2014**, *141*, 143–149. [CrossRef]

15. Ye, X.L.; Cao, D.; Song, F.Y.; Fan, G.R.; Wu, F.H. Preparative separation of nine flavonoids from Pericarpium Citri Reticulatae by preparative-HPLC and HSCCC. *Sep. Purif. Technol.* **2016**, *51*, 807–815.

16. Feng, J.T.; Xiao, Y.S.; Guo, Z.M.; Yu, D.H.; Jin, Y.; Liang, X.M. Purification of compounds from *Lignum dalbergia Odorifera* using two-dimensional preparative chromatography with Click oligo (ethylene glycol) and C18 column. *J. Sep. Sci.* **2011**, *34*, 299–307. [CrossRef] [PubMed]

17. Chen, J.Y.; He, L.H.; Yang, T. Scale-up purification for rutin hyrdrolysates by high-performance counter-current chromatography coupled with semi-preparative high-performance liquid chromatography. *Sep. Purif. Technol.* **2016**, *51*, 1523–1530. [CrossRef]

18. Zhu, L.C.; Li, H.; Liang, Y.; Wang, X.H.; Xie, H.C.; Zhang, T.Y.; Ito, Y. Application of high-speed counter-current chromatography and preparative high-performance liquid chromatography mode for rapid isolation of anthraquinones from *Morinda officinalis* How. *Sep. Purif. Technol.* **2009**, *70*, 147–152. [CrossRef]

19. Li, X.L.; Liu, Y.F.; Shen, A.J.; Wang, C.R.; Yan, J.Y.; Zhao, W.J.; Liang, X.M. Efficient purification of active bufadienolides by a class separation method based on hydrophilic solid-phase extraction and reversed-phase high performance liquid chromatography. *J. Pharmaceut. Biomed.* **2014**, *97*, 54–64. [CrossRef] [PubMed]

20. Wu, D.F.; Jiang, X.H.; Wu, S.H. Direct purification of tanshinones from *Salvia miltiorrhiza* Bunge by high-speed counter-current chromatography without presaturation of the two-phase solvent mixture. *J. Sep. Sci.* **2010**, *33*, 67–73. [CrossRef] [PubMed]

21. Li, H.B.; Chen, F. Preparative isolation and purification of six diterpenoids from the Chinese medicinal plant *Salvia miltiorrhiza* by high-speed counter-current chromatography. *J. Chromatogr. A* **2001**, *925*, 109–114. [CrossRef]

22. Tian, G.L.; Zhang, Y.B.; Zhang, T.Y.; Yang, F.Q.; Ito, Y. Separation of tanshinones from *Salvia miltiorrhiza* Bunge by high-speed counter-current chromatography using stepwise elution. *J. Chromatogr. A* **2000**, *904*, 107–111. [CrossRef]

23. Wang, J.X.; Guo, Z.M.; Shen, A.J.; Yu, L.; Xiao, Y.S.; Xue, X.Y.; Zhang, X.L.; Liang, X.M. Hydrophilic-subtraction model for the characterization and comparison of hydrophilic interaction liquid chromatography columns. *J. Chromatogr. A* **2015**, *1398*, 29–46. [CrossRef] [PubMed]

24. Jiang, L.; Tao, Y.D.; Wang, D.; Tang, C.C.; Shao, Y.; Wang, Q.L.; Zhao, X.H.; Zhang, Y.Z.; Mei, L.J. A novel two-dimensional preparative chromatography method designed for the separation of traditional animal Tibetan medicine *Osteon myospalacem* Baileyi. *J. Sep. Sci.* **2014**, *37*, 3060–3066. [CrossRef] [PubMed]

25. Park, J.Y.; Kim, J.H.; Kim, Y.M.; Jeong, H.J.; Kim, D.W.; Park, K.H.; Kwon, H.J.; Park, S.J.; Lee, W.S.; Ryu, Y.B. Tanshinones as selective and slow-binding inhibitors for SARS-CoV cysteine proteases. *Bioorg. Med. Chem.* **2012**, *20*, 5928–5935. [CrossRef] [PubMed]

26. Lin, L.Z.; Blasko, G.; Cordell, G.A. Diterpenes of *Salvia prionitis*. *Phytochemistry* **1989**, *28*, 177–181. [CrossRef]

27. Jang, T.S.; Zhang, H.; Kim, G.; Kim, D.W.; Min, B.S.; Kang, W.; Son, K.H.; Na, M.; Lee, S.H. Bioassay-guided isolation of fatty acid synthase inhibitory diterpenoids from the roots of *Salvia miltiorrhiza* Bunge. *Arch. Pharm. Res.* **2012**, *35*, 481–486. [CrossRef] [PubMed]

28. Li, W.H.; Chang, S.T.; Chang, S.C.; Chang, H.T. Isolation of antibacterial diterpenoids from *Cryptomeria japonica* bark. *Nat. Prod. Res.* **2008**, *22*, 1085–1093. [CrossRef] [PubMed]

**Sample Availability:** Samples of the tanshinone IIA, salvinolone, cryptotanshinone and ferruginol are available from the authors.

*molecules*

MDPI

*Article*

# Preconcentration and Determination of Perfluoroalkyl Substances (PFASs) in Water Samples by Bamboo Charcoal-Based Solid-Phase Extraction Prior to Liquid Chromatography–Tandem Mass Spectrometry

Ze-Hui Deng [1,2], Chuan-Ge Cheng [2], Xiao-Li Wang [2,*], Shui-He Shi [3], Ming-Lin Wang [1,*] and Ru-Song Zhao [2]

[1]  College of Food Science and Engineering, Shandong Agricultural University, Taian 271018, China; dengzh940209@126.com
[2]  Key Laboratory for Applied Technology of Sophisticated Analytical Instruments of Shandong Province, Analysis and Test Center, Qilu University of Technology (Shandong Academy of Sciences), Jinan 250014, China; chengchg@sdas.org (C.-G.C.); zhaors1976@126.com (R.-S.Z.)
[3]  Environmental Monitoring Station of Dongming Environmental Protection Bureau, Dongming 274500, China; 18354080666@163.com
*  Correspondence: mlwang@sdau.edu.cn (M.-L.W.); wxlatc@163.com (X.-L.W.)

Academic Editor: Victoria F. Samanidou
Received: 22 March 2018; Accepted: 12 April 2018; Published: 14 April 2018

**Abstract:** In this work, bamboo charcoal was used as solid-phase extraction adsorbent for the enrichment of six perfluoroalkyl acids (PFAAs) in environmental water samples before liquid chromatography–tandem mass spectrometry analysis. The specific porous structure, high specific surface area, high porosity, and stability of bamboo charcoal were characterized. Several experimental parameters which considerably affect extraction efficiency were investigated and optimized in detail. The experimental data exhibited low limits of detection (LODs) (0.01–1.15 ng/L), wide linear range (2–3 orders of magnitude and $R \geq 0.993$) within the concentration range of 0.1–1000 ng/L, and good repeatability (2.7–5.0%, $n = 5$ intraday and 4.8–8.3%, $n = 5$ interday) and reproducibility (5.3–8.0%, $n = 3$). Bamboo charcoal was successfully used for the enrichment and determination of PFAAs in real environmental water samples. The bamboo charcoal-based solid-phase extraction coupled with liquid chromatography–tandem mass spectrometry analysis possessed great potential in the determination of trace PFAA levels in environmental water samples.

**Keywords:** bamboo charcoal; solid-phase extraction; perfluoroalkyl acids; liquid chromatography–tandem mass spectrometry

## 1. Introduction

Perfluoroalkyl Substances (PFASs) consists of a C–F bond, which is one of the strongest chemical bonds; these compounds are both hydrophobic and lipophobic [1]. Fluorine is the most electronegative element, and the unique physical and chemical properties of perfluorinated organic compounds can be achieved by the introduction of fluorine atoms. Perfluorinated organic compounds, which exhibit good chemical stability, outstanding surface activity, and excellent thermal stability, are extensively applied in cutting-edge technologies, major industrial projects, pharmaceuticals, pesticides, and other industries. PFASs are distributed in various samples, such as water [2–5], soil and sediments [6–9], biological samples [10], and food samples [11] because of their high stability. The hazards of these organic compounds have been reported recently. The Organization for Economic Cooperation and Development and the US Environmental Protection Agency classified PFASs as "potentially carcinogenic substances". These compounds have attracted considerable attention worldwide. Improved methods are needed to be sought for monitoring of slow PFAS levels in a variety of samples.

Many organic contaminants have been detected at trace levels in recent years because of the coupling of gas or liquid chromatography (LC) with mass spectrometry (MS) techniques [12,13]. LC coupled with tandem MS (LC-MS/MS) is an effective analytical method for the sensitive and selective detection of PFASs [14–16]. Direct analysis of PFASs is almost impossible because of their ultra-low concentration in various samples and the complexity of sample matrices [14]. A simple, convenient, time-saving, and solvent-free sample pretreatment technique prior to LC-MS/MS analysis is required.

Sample processing techniques, including pressurized-liquid extraction (PLE) [17], solvent extraction [4], dispersive solid-phase extraction [18–20], solid-phase extraction (SPE) [21–23], magnetic solid-phase extraction (MSPE) [24–26] and other techniques, are utilized to enrich trace PFAS levels in environmental and biological samples prior to chromatographic analysis. Organic solvents used in PLE are toxic to the environment, and this technique is time consuming. In the 1970s, SPE technology has replaced traditional liquid–liquid extraction as an effective pretreatment method. SPE technology has been widely used in food, biological, pharmaceutical, and environmental analyses because of its reliability, high efficiency, simple operation, and low solvent consumption [27,28]. Traditionally, C18, oasis WAX sorbent, and HLB polymer were used as SPE sorbents to enrich PFCs in biological and environmental samples [29–32]. Bamboo charcoal, a new biomaterial with special microporous characteristics, has attracted great attention in many fields in recent years. Bamboo charcoal has been widely used for the enrichment of pollutants in environmental samples because of its relatively low price, specific porous structure, high porosity, and stability [33,34].

In this study, bamboo charcoal was used as a SPE sorbent to enrich six perfluoroalkyl acids (PFAAs) in water samples. The effects of bamboo charcoal on the experimental parameters, such as eluent, eluent flow rates, pH, sample volume, and eluent volume, were evaluated, and the parameters on extraction efficiencies were optimized. A simple, low-cost, and highly selective and sensitive SPE-HPLC-MS/MS method was established and applied for the sensitive determination of PFASs in environmental water samples.

## 2. Results and Discussion

### 2.1. Characterization of Bamboo Charcoal

A SEM micrograph of the bamboo charcoal is shown in Figure 1A, and the porous structure of the material can be seen clearly. The BET-specific surface area of the bamboo charcoal was 31.932 $m^2/g$. Bamboo charcoal can be used as an effective sorbent for environmental pollutants because of its plentiful cavity construction and high specific surface area.

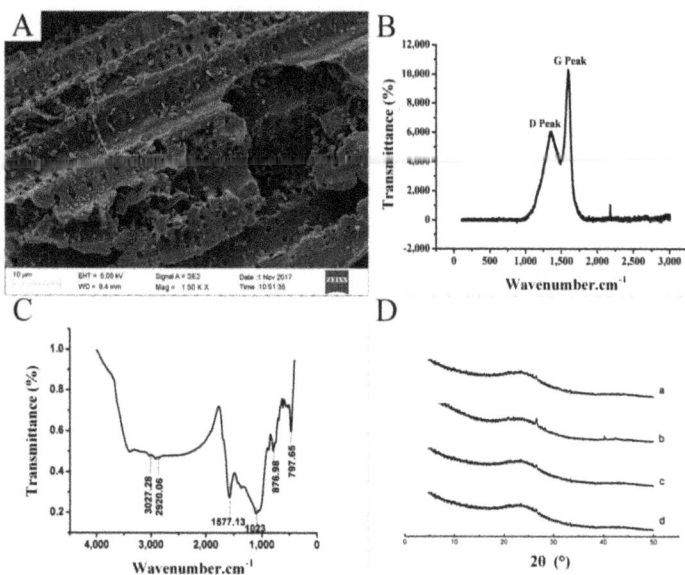

**Figure 1.** (**A**) SEM image of the bamboo charcoal at 1,500× magnification; (**B**) Raman spectra of the bamboo charcoal; (**C**) FTIR spectra of the bamboo charcoal; and (**D**) XRD patterns of the bamboo charcoal in: air (a); HCl aqueous solution, pH 2 (b); NaOH aqueous solution, pH 12 (c); and methanol for 24 h (d).

A Raman spectrum of the bamboo charcoal is shown in Figure 1B. The peak positions of D and G were determined by the mechanical constants of C–C bonds in the carbon network plane of the graphite microcrystal or graphite-like microcrystal. Various oxygen-containing functional groups were present at the edge of the graphite-like microcrystal that formed in the low-temperature carbonization stage of biomass carbon. Ether bonds may also be present between the monolayer carbon planes of the graphite-like microcrystals. The existence of these functional groups or bonds may affect the delocalized π electron behavior in the carbon network plane. Thus, the mechanical constants of the C–C bond were increased or decreased, and Raman shifts were detected. The D peak was caused by the sp2-hybridized-carbon atoms at the edge of the graphite microcrystal, and the G peak was caused by the translational vector of the symmetrical structure in the carbon network plane of the graphite microcrystal. Thus, the oxygen-containing functional groups between the carbon network and those at the edge of the carbon network exhibited different effects on their Raman spectra.

As shown in the FTIR spectra (Figure 1C), the bamboo charcoal exhibited an absorbance peak at approximately 1577.13 $cm^{-1}$ due to the stretching vibration of the carbonyl group (C=O). Two absorbance peaks at approximately 797.65 and 1023 $cm^{-1}$ were assigned, respectively, to the out-of-plane bending vibration of the C–H group and stretching vibrations of the C–O and C–O–O–C groups. These results verified the existence of a carbonyl group (C=O) on the bamboo charcoal. Considering that the electronegativity of an oxygen atom (3.5) is higher than that of a carbon atom (2.5), the electron cloud distribution of the C=O bond is biased toward the oxygen atom, which determines polarity and chemical reactivity with numerous polar substances of the C=O group. PFCs are a group of environmental organic pollutants with strong polarity, and the C=O group on the bamboo charcoal can react with PFCs, supporting bamboo charcoal as a novel SPE sorbent for sensitive PFC extractionin environmental water samples.

The chemical stability of the bamboo charcoal in extreme conditions, such as acidic, alkaline, and organic solvents, was investigated in this study. The bamboo charcoal (500 mg) was immersed separately in NaOH solution (pH = 13), HCl solution (pH = 2), and methanol at room temperature for 24 h. As shown in Figure 1D, no evident changes in the XRD patterns were observed under different experimental conditions. These results indicated that bamboo charcoal is stable in aqueous solution with a broad pH range of 2–13 and organic solvents and is suitable for environmental pollutant analysis.

## 2.2. Optimization of the Experimental Parameters

To acquire optimized extraction conditions, effective parameters, such as the eluent, eluent volume and flow rate, sample pH, and sample volume and flow rate, were investigated and optimized in detail. In this work, 100 mL of ultrapure water spiked with 10 μg/LPFAAs (PFHxS, PFHpA, PFOA, PFOS, PFNA, and PFDA) was used to investigate the SPE performance of bamboo charcoal.

The eluent is one of the most important factors in sample preconcentration procedure. In this experiment, the solvents acetone, methanol, acetonitrile, dichloromethane, and *n*-hexane were tested. The desorption efficiency of these five solvents are shown in Figure 2A. Acetone exhibited the best elution performance for the PFAAs among the five studied solvents. Therefore, acetone was chosen as the desorption solvent in subsequent work.

The influence of eluent (acetone) volume (2–14 mL) on the desorption efficiency of PFAAs was examined (Figure 2B). The desorption efficiencies increased as the eluent volume increased from 2 mL to 12 mL. The desorption efficiency did not significantly increase at >12 mL elution volumes. Thus, 12 mL of acetone as the eluent volume was adopted in the following experiments.

Eluent flow rate is also an important factor that affects desorption efficiency because it influences the contact time between the molecules of target pollutants and the eluent [35]. The eluent flow rate was investigated and optimized at 0.5, 1, 2, and 3 mL/min to save desorption time and obtain satisfactory results (Figure 2C). The recoveries of the six PFAAs increased with decreasing flow rate. Thus, the flow rate of 0.5 mL/min was chosen for subsequent analytical experiments.

The influence of pH (2.0–12.0) on the extraction efficiency was investigated. The recoveries reached the optimal level at pH 4.0 (Figure 2D). These results illustrated that PFAAs can be effectively adsorbed onto the bamboo charcoal sorbents under acidic conditions. The possible reason is that the pH of the sample solution affects the forms of PFAAs existing in the aqueous samples. Under acidic conditions, PFAAs mainly exist as unionized acid, whereas, under neutral or alkaline conditions, PFAAs are mainly ionized and soluble in water samples, leading to decreased adsorptive efficiency from water samples to adsorbents [27]. Thus, the sample pH was adjusted to 4.0 in subsequent experiments.

The flow rate of 2–5 mL/min was investigated to save analytical time and obtain satisfactory experimental results (Figure 2E). At 2–5 mL/min flow rates, the recoveries obtained were between 81.39% and 99.12%. The sample flow rate of 5 mL/min was selected for the subsequent experiments of the PFAAs. Sample volume was also optimized in the experiment. The recovery remained stable when the sample volume increased from 100 mL to 1000 mL (Figure 2F), and the reactions between the targeted pollutants and bamboo charcoal were not affected by the sample volume. The sample volume of 100 mL was selected for the subsequent experiments.

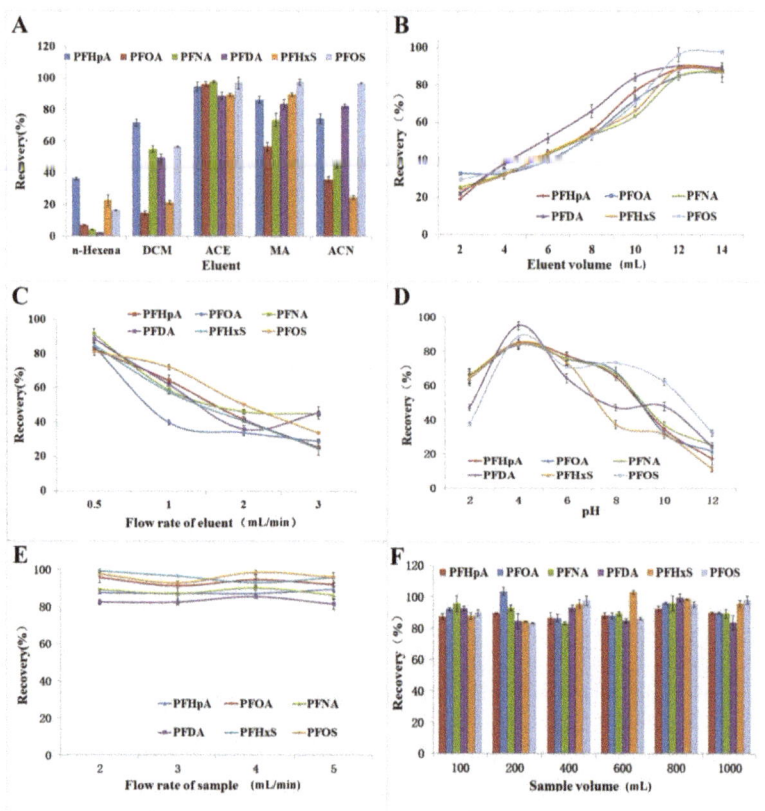

**Figure 2.** Effects of the: eluent (**A**); eluent volume (**B**); flow rate of eluent (**C**); pH (**D**); flow rate of sample (**E**); and sample volume (**F**) on the recoveries of the six PFAAs. The PFAA concentration in the water samples was 100 ng/L.

### 2.3. Method Evaluation

The analytical data for the six kinds of PFAAs using bamboo charcoal as SPE adsorbent under optimal parameters are summarized in Table 1. The developed method exhibited good linearity (R ≥ 0.993) within the concentration range of 0.1–1000 ng/L. The limits of detection (LODs) based on signal-to-noise (S/N) ratios of 3 ranged from 0.01 ng/L to 1.15 ng/L. The limits of quantification (LOQs), which is calculated by S/N ratios of 10, ranged from 0.03 ng/L to 3.85 ng/L. The relative standard deviations (RSDs) of the intraday (*n* = 5) and interday (*n* = 5) experiments when using bamboo charcoal as the SPE adsorbent coupled to LC-MS/MS were in the range of 2.7–5.0% and 4.8–8.3%, respectively, for the six kinds of PFAAs. This finding illustrated the good repeatability and reproducibility of this method by using a single SPE column. Three SPE columns were prepared under the same conditions, and the column-to-column reproducibility (*n* = 3) was 5.3–8.0% for the six PFAAs (100 ng/L). As shown in Table 2, this method produced a wider linear range, lower LODs and LOQs, and higher accuracy efficiency compared with other methods mentioned in previous studies [20,24,31,32,36]. Moreover, one bamboo charcoal column can be reused more than 10 times without a detectable extraction efficiency loss. The experimental data exhibited that bamboo charcoal is suitable as a novel extraction adsorbent for the analysis of strong polar PFAAs in environmental samples.

**Table 1.** Analytical data of the SPE method.

| Compounds | Linear Range (ng/L) | R | LODs (ng/L) | LOQs (ng/L) | Repeatability (%, n = 5) | | Column-to-Column Reproducibility (%, n = 3) |
|---|---|---|---|---|---|---|---|
| | | | | | Intraday | Interday | |
| PFHpA | 1.0–200.0 | 0.999 | 0.11 | 0.37 | 3.3 | 6.8 | 6.4 |
| PFOA | 1.0–200.0 | 0.999 | 0.07 | 0.22 | 2.7 | 5.4 | 7.3 |
| PFNA | 4.0–1000 | 0.999 | 1.15 | 3.85 | 3.6 | 4.8 | 7.8 |
| PFDA | 10.0–1000 | 0.997 | 0.88 | 3.68 | 4.1 | 8.3 | 8.0 |
| PFHxS | 0.1–100 | 0.993 | 0.01 | 0.03 | 5.0 | 5.1 | 5.8 |
| PFOS | 0.1–100 | 0.998 | 0.01 | 0.03 | 2.9 | 7.0 | 5.3 |

**Table 2.** Method comparisons for the analysis of the six PFAAs.

| Material | Analytical Methods | Linear Range (ng/L) | LODs (ng/L) | RSD (%) | Recoveries (%) | References |
|---|---|---|---|---|---|---|
| $Fe_3O_4$@mSiO$_2$-F17 | MSPE-HPLC-MS/MS | 250–1,000,000 | 20–50 | 2.6–14.2 | 83.13–92.42 | [24] |
| C18, PSA, GCB | QuEChERS-HPLC-MS/MS | 100–10,000 | 50–200 | 2.1–11.9 | 70.3–108.1 | [20] |
| HLB | SPE-HPLC-MS | 500–200,000 | 150–900 | 7.5–11.8 | 73–88 | [31] |
| CTAB-MCM-41 | μ-SPE-LC-MS | 1000–100,000 | 970–2700 | 5.4–13.5 | 77–120 | [32] |
| Octadecylsiyl particles | SPE-Reversed Phase-HPLC-MS | - | 25 | 0.5–10.8 | 79.2–96.1 | [36] |
| Bamboo charcoal | SPE-LC-MS/MS | 0.1–1250 | 0.01–1.44 | 0.4–8.3 | 86.9–117.2 | This work |

PSA: *N*-propylethylendiamine; GCB: graphitized carbon blacks; HLB: The HLB adsorbent is a macroporous copolymer that is polymerized from lipophilic divinylbenzene and hydrophilic *N*-vinylpyrrolidone in a certa proportion; CTAB-MCM-41: a kind of new material (cetyltrimethylammonium bromide contained MCM-41); MSPE: magnetic solid phase extraction; QuEChERS: a quick, easy, cheap, effective, rugged and safe sample pretreatment method.

## 2.4. Analysis of Fortified Samples for Recoveries Calculation

The proposed SPE method with bamboo charcoal as adsorbent was then applied to analyze PFAAs in four real water samples, namely, barreled drinking water, tap water, pond water, and water collected from Dagu Port Scenic Resort. As shown in Table 3, PFHxS was detected at 0.56 ng/L in the tap water samples, and both PFHxS and PFOA were detected at 4.61 and 3.93 ng/L in pond water samples, respectively. No PFAA pollutants were detected in the barreled drinking water samples and water samples collected from Dagu Port Scenic Resort. Recovery testing was performed by spiking three different levels of PFAAs (20, 50, and 100 ng/L) in the four samples. The recoveries were within 86.9–117.2% at 0.4–8.3% RSDs. Typical chromatograms of PFAAs in an environmental water sample are illustrated in Figure 3. We can conclude from all of the experimental data that the analytical method established in this work is suitable for the analysis of PFAAs at trace levels in real water samples.

**Table 3.** Analytical results for the determination of the six PFAAs in real water samples.

| Samples | Added (ng/L) | PFHpA | PFOA | PFNA | PFDA | PFHxS | PFOS |
|---|---|---|---|---|---|---|---|
| Barreled drinking water | 0.0 | ND [a] | ND [a] | ND [a] | ND [a] | ND [a] | ND [a] |
| | 20.0 | 104.2 [b] ± 6.8 [c] | 102.8 ± 3.1 | 97.2 ± 1.3 | 102.4 ± 2.3 | 96.1 ± 4.1 | 92.0 ± 2.4 |
| | 50.0 | 94.3 ± 4.7 | 104.9 ± 4.2 | 96.0 ± 3.2 | 109.3 ± 1.8 | 100.5 ± 5.5 | 100.5 ± 3.1 |
| | 100.0 | 89.7 ± 7.0 | 99.7 ± 5.2 | 99.2 ± 1.6 | 100.3 ± 3.7 | 103.2 ± 3.9 | 96.8 ± 5.1 |
| Tap water | 0.0 | ND [a] | ND [a] | ND [a] | ND [a] | 0.56 | ND [a] |
| | 20.0 | 95.4 ± 0.9 | 87.5 ± 6.3 | 90.6 ± 4.1 | 99.4 ± 8.3 | 99.1 ± 6.1 | 89.3 ± 2.9 |
| | 50.0 | 98.6 ± 1.4 | 94.6 ± 2.7 | 93.8 ± 2.3 | 95.3 ± 3.8 | 94.5 ± 5.1 | 93.2 ± 3.0 |
| | 100.0 | 111.4 ± 5.3 | 93.7 ± 2.3 | 98.5 ± 3.2 | 98.7 ± 6.2 | 91.2 ± 1.7 | 91.6 ± 1.4 |
| Pond water | 0.0 | ND [a] | 3.93 | ND [a] | ND [a] | 4.61 | ND [a] |
| | 20.0 | 92.8 ± 6.2 | 117.2 ± 3.8 | 103.7 ± 3.1 | 86.4 ± 0.9 | 83.4 ± 4.6 | 107.3 ± 6.9 |
| | 50.0 | 98.1 ± 7.4 | 105.6 ± 4.5 | 104.2 ± 2.4 | 89.3 ± 4.2 | 86.9 ± 3.1 | 105.3 ± 2.5 |
| | 100.0 | 107.3 ± 2.4 | 102.2 ± 4.1 | 101.3 ± 5.1 | 91.0 ± 1.7 | 84.1 ± 1.9 | 99.6 ± 3.4 |
| Port water | 0.0 | ND [a] | ND [a] | ND [a] | ND [a] | ND [a] | ND [a] |
| | 20.0 | 92.4 ± 1.4 | 98.4 ± 6.1 | 101.1 ± 0.4 | 85.4 ± 4.2 | 97.3 ± 0.4 | 93.2 ± 4.1 |
| | 50.0 | 87.3 ± 4.3 | 99.1 ± 4.3 | 100.3 ± 3.6 | 87.5 ± 1.3 | 91.4 ± 4.7 | 97.5 ± 7.3 |
| | 100.0 | 102.4 ± 5.1 | 92.9 ± 5.5 | 97.8 ± 6.7 | 93.6 ± 1.7 | 89.6 ± 2.9 | 91.4 ± 1.8 |

[a] Not detected; [b] Mean value of three determinations; [c] Standard deviation (*n* = 3).

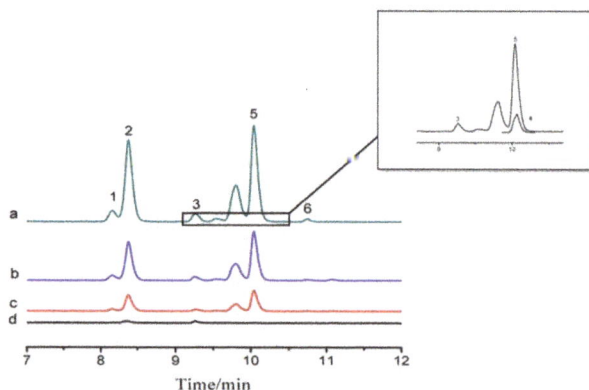

**Figure 3.** Typical chromatograms of the six PFAAs in real water samples. Pond water spiked at: (a) 100; (b) 50; and 20 ng/L (c); and pond water (d). (1) PFHpA; (2) PFHxS; (3) PFOA; (4) PFNA; (5) PFOS; and (6) PFDA.

## 3. Materials and Methods

### 3.1. Chemicals and Reagents

Bamboo charcoal was purchased from Zhejiang Forasen Bamboo Tec Co., Ltd. (Zhejiang, China). The bamboo charcoal was first triturated in a glass mortar, sieved through an 80-mesh sieve, and dried at 80 °C for 2 h [27].

Perfluorohexanesulfonate (PFHxS), perfluoroheptanoic acid (PFHpA), perfluorooctanoic acid (PFOA), perfluorooctanesulfonic acid (PFOS), perfluorononanoic acid (PFNA), and perfluorodecanoic acid (PFDA) were purchased from AnpuShiyan Tech Co., Ltd. (Shanghai, China). Methanol, acetone, and acetonitrile were obtained from Tedia Company (Fairfield, OH, USA). Dichloromethane and *n*-hexane were purchased from Concord Technology (Tianjin, China). All other reagents and chemicals used in this experiment were of at least analytical grade. PFAS stock solution containing PFOA, PFHpA, PFNA, PFDA, PFHxS, and PFOS at 1 µg/mL was prepared by dissolving 0.1 mg of each of the six types of PFCs in a 100 mL volumetric flask. A series of standard solutions was obtained by gradually diluting the stock solution with methanol. All solutions were stored at 4 °C in the dark prior to use.

### 3.2. Instrument

In this work, a Thermo Ultimate 3000 Liquid Chromatograph (Thermo Scientific, Waltham, MA, USA) coupled with an AB SCIEX QTRAP 5500 triple quadrupole mass spectrometer (SCIEX, Framingham, MA, USA) was used. An Agilent XDB-C18 column (2.1 mm × 150 mm, 3.5 µm, Santa Clara, CA, USA) was used for the chromatographic separation at 40 °C. The mobile phases were 5 mmol/L NH4Ac (A) and methanol (B). The gradient elution during the chromatographic run was as follows: 0–1.0 min, 10% B; 1.1–1.5 min, 10–40% B; 1.6–12.0 min, 40–95% B; 12.1–13.0 min, 95% B; and 13.1–17 min, 10% B. The flow rate of the mobile phase was set at 0.4 mL/min, and the injection volume was 10 µL. The mass spectrometer analysis was conducted in the negative ionization mode with multiple reaction monitoring mode. The source temperature was 550 °C, and nitrogen was used as the collision gas. The ion spray voltage was −4500 V, the curtain gas was 40 psi, and the ion source gases 1 and 2 were 55 and 60 psi, respectively. The optimized MS/MS parameters are listed in Table 4.

Table 4. HPLC–MS/MS parameters for MRM acquisition of PFAAs.

| Compounds | Retention Time (min) | Precursorion (*m/z*) | Product Ion (*m/z*) | Declustering Potential (V) | Collision Energy (eV) |
|---|---|---|---|---|---|
| PFHpA | 8.29 | 363 | 319, 169 | −30, −30 | 14, 24 |
| PFOA | 9.25 | 413 | 369, 169 | −40, −30 | 14, 24 |
| PFNA | 10.07 | 463 | 419, 219 | −35, −35 | 16, 24 |
| PFDA | 10.75 | 513 | 469, 219 | −40, −40 | 18, 26 |
| PFHxS | 8.38 | 399 | 79.9, 99 | −90, −90 | 88, 72 |
| PFOS | 10.04 | 499 | 79.9, 99 | −105, −105 | 106, 98 |

The scanning electron microscopy (SEM) images of the bamboo charcoal were obtained using SUPPA™ 55 (Zeiss, Oberkochen, Germany). X-ray diffraction (XRD) measurements with the angle ranging from 10° to 50° were obtained with Cu Ka radiation on a D/max-Rbdiffractometer (Rigaku, Japan). The Brunauer–Emmett–Teller (BET)-specific surface areas of the bamboo charcoal were measured using an ASAP 2020 porosimeter (Micromeritics, Norcross, GA, USA). FTIR spectra were obtained using a Nicolet 710 IR spectrometer (Thermo Scientific, Waltham, MA, USA). The Raman spectrum of the bamboo charcoal was obtained using Renishaw inVia microscopes and a Raman spectrometer (Renishaw, Sheffield, UK).

*3.3. SPE*

Bamboo charcoal-packed cartridges were prepared based on previous literature [28,37]. Bamboo charcoal powder (300 mg) that was treated as mentioned above was packed in an empty SPE cartridge. The polypropylene frit was reset to hold the bamboo charcoal powder in place. The inlet of the cartridge was connected to a PTFE suction tube, which was inserted into the sample solution. The outlet of the cartridge was connected to a vacuum pump. The SPE cartridge was washed with purified water and acetone several times before its first use to reduce possible contaminants.

The bamboo charcoal column was washed and activated with 5 mL of purified water and 5 mL of acetone. Subsequently, 100 mL of water sample spiked with six PFAAs was passed through the pretreated cartridge at 5 mL/min. The cartridge was then rinsed with 10 mL of purified water to remove possible adsorbed matrix materials from the column. The bamboo charcoal column was then dried at negative pressure for 5 min. Subsequently, the target compounds retained on the bamboo charcoal were eluted with 12 mL of acetone, and the eluent was dried at 40 °C under nitrogen. Finally, the residue was dissolved in 1.0 mL of methanol prior to HPLC-MS/MS analysis.

*3.4. Water Sample Collection*

Four kinds of water samples, namely, barreled drinking water, tap water, pond water, and water collected from Dagu Port Scenic Resort, were used to evaluate the feasibility of the developed method. The barreled drinking water samples were obtained from the local supermarket (Jinan, China). The tap water samples were collected from our laboratory (Jinan, China). The pond water samples were acquired from the pond located at the Analysis and Test Centre (Jinan, China). After filtration through a 0.45 μm membrane filter, these water samples were stored in brown glass bottles at 4 °C for subsequent SPE extractions.

## 4. Conclusions

In this study, bamboo charcoal was used as an SPE adsorbent for the first time to enrich and analyze six kinds of new persistent organic pollutant perfluorooctanoic acids at trace levels in water samples. This novel adsorbent achieved good chemical stability; high repeatability, good reproducibility, and extraction efficiency; wide linear range (2–3 orders of magnitude); and low LODs (0.01–1.15 ng/L) for the analysis of PFAAs. An affordable and easily available material, bamboo charcoal is suitable as an SPE adsorbent for the extraction and analysis of polar organic pollutants in environmental water samples.

**Acknowledgments:** This study was supported by the National Natural Science Foundation of China (21477068 and 21407099), the Natural Science Foundation of Shandong Province (ZR2015YL003), and the Key Research and Development Program of Shandong Province (2017GSF220017 and 2015GSF117011).

**Author Contributions:** Ze-Hui Deng carried out the experiments and data acquisition; Chuan-Ge Cheng carried out data analysis and provided technical assistance; Xiao-Li Wang collated experimental data and revised the manuscript; Shui-He Shi provided the water samples; Ming-Lin Wang designed the experiments and carried out data analysis; Ru-Song Zhao analyzed experimental data and wrote the manuscript.

**Conflicts of Interest:** The authors declare no conflict of interest.

## References

1. Richardson, S.D. Environmental mass spectrometry: Emerging contaminants and current issues. *Anal. Chem.* **2012**, *84*, 747–778. [CrossRef] [PubMed]
2. De Silva, A.O.; Muir, D.C.; Mabury, S.A. Distribution of perfluorocarboxylate isomers in select samples from the North American environment. *Environ. Toxicol. Chem.* **2009**, *28*, 1801–1814. [CrossRef] [PubMed]
3. Takazawa, Y.; Nishino, T.; Sasaki, Y.; Yamashita, H.; Suzuki, N.; Tanabe, K.; Shibata, Y. Occurrence and distribution of perfluorooctanesulfonate and perfluorooctanoic acid in the rivers of Tokyo. *Water Air Soil Pollut.* **2009**, *202*, 57–67. [CrossRef]
4. Llorca, M.; Farré, M.; Picó, Y.; Barceló, D. Analysis of perfluorinated compounds insewage sludge by pressurized solvent extraction followed by liquid chromatography–mass spectrometry. *J. Chromatogr. A* **2011**, *1218*, 4840–4846. [CrossRef] [PubMed]
5. Yan, Z.; Cai, Y.; Zhu, G.; Yuan, J.; Tu, L.; Chen, C.; Yao, S. Synthesis of 3-fluorobenzoyl chloride functionalized magnetic sorbent for highly effcient enrichment of perfluorinated compounds from river water samples. *J. Chromatogr. A* **2013**, *1321*, 21–29. [CrossRef] [PubMed]
6. Higgins, C.P.; Field, J.A.; Criddle, C.S.; Luthy, R.G. Quantitative determination of perfluorochemicals in sediments and domestic sludge. *Environ. Sci. Technol.* **2005**, *39*, 3946–3956. [CrossRef] [PubMed]
7. Washington, J.W.; Henderson, W.M.; Ellington, J.J.; Jenkins, T.M.; Evans, J.J. Analysis of perfluorinated carboxylic acids in soils II: Optimization of chromatography and extraction. *J. Chromatogr. A* **2008**, *1181*, 21–32. [CrossRef] [PubMed]
8. Nakata, H.; Kannan, K.; Nasu, T.; Cho, H.S.; Sinclair, E.; Takemurai, A. Perfluorinated contaminants in sediments and aquatic organisms collected from shallow water and tidal flat areas of the Ariake Sea, Japan: Environmental fate of perfluorooctanesulfonate in aquatic ecosystems. *Environ. Sci. Technol.* **2006**, *40*, 4916–4921. [CrossRef] [PubMed]
9. Alzaga, R.; Salgado-Petinal, C.; Jover, E.; Bayona, J.M. Development of a procedure for the determination of perfluorocarboxylic acids in sediments by pressurised fluid extraction, headspace solid-phase microextraction followed by gas chromatographic-mass spectrometric determination. *J. Chromatogr. A* **2005**, *1083*, 1–6. [CrossRef] [PubMed]
10. Villaverde-de-Sáa, E.; Quintana, J.B.; Rodil, R.; Ferrero-Refojos, R.; Rubí, E.; Cela, R. Determination of perfluorinated compounds in mollusks by matrix solid-phase dispersion and liquid chromatography-tandem mass spectrometry. *Anal. Bioanal. Chem.* **2012**, *402*, 509–518. [CrossRef] [PubMed]
11. Ballesteros-Gómez, A.; Rubio, S.; van Leeuwen, S. Tetrahydrofuran-water extraction, in-line clean-up and selective liquid chromatography/tandem mass spectrometry for the quantitation of perfluorinated compounds in food at the low picogram per gram level. *J. Chromatogr. A* **2010**, *1217*, 5913–5921. [CrossRef] [PubMed]
12. Farré, M.; Kantiani, L.; Petrovic, M.; Pérez, S.; Barceló, D. Achievements and future trends in the analysis of emerging organic contaminants in environmental samples by mass spectrometry and bioanalytical techniques. *J. Chromatogr. A* **2012**, *1259*, 86–99. [CrossRef] [PubMed]
13. Farre', M.; Barcelo', D. Analysis of emerging contaminants in food. *Trends Anal. Chem.* **2013**, *43*, 240–253. [CrossRef]

14. Wang, X.; Zhang, Y.; Li, F.W.; Zhao, R.S. Carboxylated carbon nanospheres as solid-phase extraction adsorbents for the determination of perfluorinated compounds in water samples by liquid chromatography–tandem mass spectrometry. *Talanta* **2018**, *178*, 129–133. [CrossRef] [PubMed]

15. García-Valcárcel, A.I.; Tadeo, J.L. Fast ultrasound-assisted extraction combined with LC–MS/MS of perfluorinated compounds in manure. *J. Sep. Sci.* **2013**, *36*, 2507–2513. [CrossRef] [PubMed]

16. Zhang, J.; Wan, Y.; Li, Y.; Zhang, Q.; Xu, S.; Zhu, H.; Shu, B. A rapid and high-throughput quantum dots bioassay for monitoring of perfluorooctanesulfonate in environmental water samples. *Environ. Pollut.* **2011**, *159*, 1348–1353. [CrossRef] [PubMed]

17. González-Barreiro, C.; Martínez-Carballo, E.; Sitka, A.; Scharf, S.; Gans, O. Method optimization for determination of selected perfluorinated alkylated substances in water samples. *Anal. Bioanal. Chem.* **2006**, *386*, 2123–2132. [CrossRef] [PubMed]

18. Yang, L.; Jin, F.; Zhang, P.; Zhang, Y.; Wang, J.; Shao, H.; Jin, M.; Wang, S.; Zheng, L.; Wang, J. Simultaneous determination of perfluorinated compounds in edible oil by gel-permeation chromatography combined with dispersive solid-phase extraction and liquid chromatography–tandem mass spectrometry. *J. Agric. Food Chem.* **2015**, *63*, 8364–8371. [CrossRef] [PubMed]

19. Surma, M.; Wiczkowski, W.; Cieślik, E.; Zieliski, H. Method development for the determination of PFOA and PFOS in honey based on the dispersive Solid Phase Extraction (*d*-SPE) with micro-UHPLC–MS/MS system. *Microchem. J.* **2015**, *121*, 150–156. [CrossRef]

20. He, J.L.; Peng, T.; Xie, J.; Dai, H.H.; Chen, D.D.; Yue, Z.F.; Fan, C.L.; Li, C. Determination of 20 Perfluorinated Compounds in Animal Liver by HPLC-MS/MS. *Chin. J. Anal. Chem.* **2015**, *43*, 40–48. [CrossRef]

21. Sun, Z.; Zhang, C.; Yan, H.; Han, C.; Chen, L.; Meng, X.; Zhou, Q. Spatiotemporal distribution and potential sources of pefluoroalkyl acids in Huangpu River, Shanghai, China. *Chemosphere* **2017**, *174*, 127–135. [CrossRef] [PubMed]

22. Enevoldsen, R.; Juhler, R.K. Perfluorinated compounds (PFCs) in groundwater and aqueous soil extracts: Using inline SPE-LC-MS/MS for screening and sorption characterisation of perfluorooctanesulphonate and related compounds. *Anal. Bioanal. Chem.* **2010**, *398*, 1161–1172. [CrossRef] [PubMed]

23. Zhu, P.; Ling, X.; Liu, W.; Kong, L.; Yao, Y. Simple and fast determination of perfluorinated compounds in Taihu Lake by SPE-UHPLC–MS/MS. *J. Chromatogr. B Analyt. Technol. Biomed. Life Sci.* **2016**, *1031*, 61–67. [CrossRef] [PubMed]

24. Liu, X.; Yu, Y.; Li, Y.; Zhang, H.; Ling, J.; Sun, X.; Feng, J.; Duan, G. Fluorocarbon-bonded magnetic mesoporous microspheres for the analysis of perfluorinated compounds in human serum by high-performance liquid chromatography coupled to tandem mass spectrometry. *Anal. Chim. Acta* **2014**, *844*, 35–43. [CrossRef] [PubMed]

25. Zhou, Y.; Tao, Y.; Li, H.; Zhou, T.; Jing, T.; Zhou, Y.; Mei, S. Occurrence investigation of perfluorinatedcompounds in surface water from East Lake (Wuhan, China) upon rapid and selective magnetic solid-phase extraction. *Sci. Rep.* **2016**, *6*, 38633. [CrossRef] [PubMed]

26. Ma, Y.R.; Zhang, X.L.; Zeng, T.; Cao, D.; Zhou, Z.; Li, W.H.; Niu, H.; Cai, Y.Q. Polydopamine-Coated Magnetic Nanoparticles for Enrichment and Direct Detection of Small Molecule Pollutants Coupled with MALD-ITOF-MS. *ACS Appl. Mater. Interfaces* **2013**, *5*, 1024–1030. [CrossRef] [PubMed]

27. Zhao, R.S.; Wang, X.; Wang, X.; Lin, J.M.; Yuan, J.P.; Chen, L.Z. Using bamboo charcoal as solid-phase extraction adsorbent for the ultratrace-level determination of perfluorooctanoic acid in water samples by high-performance liquid chromatography-mass spectrometry. *Anal. Bioanal. Chem* **2008**, *390*, 1671–1676. [CrossRef] [PubMed]

28. Zhao, R.S.; Wang, X.; Yuan, J.P.; Lin, J.M. Investigation of feasibility of bamboo charcoal as solid-phase extraction adsorbent for the enrichment and determination of four phthalate esters in environmental water samples. *J. Chromatogr. A* **2008**, *1183*, 15–20. [CrossRef] [PubMed]

29. Boone, J.S.; Guan, B.; Vigo, C.; Boone, T.; Byrne, C.; Ferrario, J. A method for the analysis of perfluorinated compounds in environmental and drinking waters and the determination of their lowest concentration minimal reporting levels. *J. Chromatogr. A* **2014**, *1345*, 68–77. [CrossRef] [PubMed]

30. Cao, D.; Wang, Z.; Han, C.; Cui, L.; Hu, M.; Wu, J.; Liu, Y.; Cai, Y.; Wang, H.; Kang, Y. Quantitative detection of trace perfluorinated compounds in environmental water samples by Matrix-assisted Laser Desorption/Ionization-Time of Flight Mass Spectrometry with 1,8-bis(tetramethylguanidino)-naphthalene as matrix. *Talanta* **2011**, *85*, 345–352. [CrossRef] [PubMed]

31. Teng, J.W.; Tang, S.Z.; Ou, S.Y. Determination of perfluorooctanesulfonate and perfluorooctanoate in water samples by SPE-HPLC/electrospray ion trap mass spectrometry. *Microchem. J.* **2009**, *93*, 55–59. [CrossRef]

32. Lashgari, M.; Lee, H.K. Determination of perfluorinated carboxylic acids in fish fillet by micro-solid phase extraction, followed by liquid chromatography–triple quadrupole mass spectrometry. *J. Chromatogr. A* **2014**, *1369*, 26–32. [CrossRef] [PubMed]

33. Zhao, R.S.; Yuan, J.P.; Jiang, T.; Shi, J.B.; Cheng, C.G. Application of bamboo charcoal as solid-phase extraction adsorbent for the determination of atrazine and simazine in environmental water samples by high-performance liquid chromatography-ultra violet detector. *Talanta* **2008**, *76*, 956–959. [CrossRef] [PubMed]

34. Zhao, R.S.; Wang, X.; Yuan, J.P.; Wang, X.D. Sensitive determination of phenols in environmental water samples with SPE packed with bamboo carbon prior to HPLC. *J. Sep. Sci.* **2009**, *32*, 630–636. [CrossRef] [PubMed]

35. Zhou, Q.; Xiao, J.; Wang, W. Using multi-walled carbon nanotubes as solid phase extraction adsorbents to determine dichlorodiphenyltrichloroethane and its metaboliteat trace level in water samples by high performance liquid chromatography with UV detection. *J. Chromatogr. A* **2006**, *1125*, 152–158. [CrossRef] [PubMed]

36. Dolman, S.; Pelzing, M. An optimized method for the determination of perfluorooctanoic acid, perfluorooctanesulfonate and other perfluorochemicals in different matrices using liquid chromatography/ion-trap mass spectrometry. *J. Chromatogr. B Analyt. Technol. Biomed. Life Sci.* **2011**, *879*, 2043–2050. [CrossRef] [PubMed]

37. Lashgari, M.; Basheer, C.; Kee Lee, H. Application of surfactant-templated ordered mesoporous material as sorbent in micro-solid phase extraction followed by liquid chromatography–triple quadrupole mass spectrometry for determination of perfluorinated carboxylic acids in aqueous media. *Talanta* **2015**, *141*, 200–206. [CrossRef] [PubMed]

**Sample Availability:** Samples of the compounds bamboo charcoal are available from the authors.

molecules

MDPI

*Article*

# Carbon Nanohorn Suprastructures on a Paper Support as a Sorptive Phase

**Julia Ríos-Gómez, Beatriz Fresco-Cala, María Teresa García-Valverde, Rafael Lucena \* and Soledad Cárdenas**

Departamento de Química Analítica, Instituto Universitario de Investigación en Química Fina y Nanoquímica IUIQFN, Universidad de Córdoba, Campus de Rabanales, Edificio Marie Curie (anexo), E-14071 Córdoba, Spain; juliariosgomez@hotmail.com (J.R.-G.); q72frcab@uco.es (B.F.-C.); q72gavam@uco.es (M.T.G.-V.); scardenas@uco.es (S.C.)
\* Correspondence: rafael.lucena@uco.es; Tel.: +34-957-211-066

Academic Editor: Victoria F. Samanidou
Received: 3 May 2018; Accepted: 22 May 2018; Published: 24 May 2018

**Abstract:** This article describes a method for the modification of paper with single-wall carbon nanohorns (SWCNHs) to form stable suprastructures. The SWCNHs form stable dahlia-like aggregates in solution that are then self-assembled into superior structures if the solvent is evaporated. Dipping paper sections into a dispersion of SWCNHs leads to the formation of a thin film that can be used for microextraction purposes. The coated paper can be easily handled with a simple pipette tip, paving the way for disposable extraction units. As a proof of concept, the extraction of antidepressants from urine and their determination by direct infusion mass spectrometry is studied. Limits of detection (LODs) were 10 ng/L for desipramine, amitriptyline, and mianserin, while the precision, expressed as a relative standard deviation, was 7.2%, 7.3%, and 9.8%, respectively.

**Keywords:** carbon nanohorns; sorptive phase; paper; microextraction; antidepressants

## 1. Introduction

Solid-phase microextraction (SPME) is a consolidated sample treatment technique that combines isolation, preconcentration, and sample introduction into one step [1]. This miniaturized technique, which can easily be automated, is based on the distribution of the analytes between the sample and the fiber coating. In this context, the reversible chemical interactions between the analyte and the sorptive phase are of paramount importance to define the efficiency and selectivity of the microextraction. SPME is in a continuous development following several evident tendencies like the development of new coatings [2,3] or the direct coupling with instrumental techniques like mass-spectrometry (MS) [4,5]. All these trends make SPME the predominant technology for microextraction.

The adaptation of the SPME principles to a specific field such as environmental analysis drove the development of thin film microextraction (TFME) [6]. Although both techniques share the same foundations, they differ in their application. TFME uses a thin sheet of a polymeric phase as a sorptive phase that may adopt several shapes [7] like a flat membrane [8] or a coated blade [9]. These formats present a higher extraction capacity compared to traditional fibers due to an increased surface to volume ratio, which has positive connotations for the thermodynamic and kinetics. Also, TFME can be automated to allow the simultaneous extraction of several samples, thus increasing the sample throughput [10]. In the typical procedure, the thin film is immersed in the sample, which is stirred to favor the analyte transference to the sorptive phase [11]. However, the thin film can also be stirred into the sample. The use of planar sorptive phases integrated into stirring units has allowed for the development of new techniques like rotating disk sorptive extraction [12] and stir membrane extraction [13,14].

As with SPME, the development of new sorptive phases is crucial in TFME. Fortunately, there is a wide range of materials that can be used, from commercial membranes to lab-made materials. Among the latter, fabric phases and electrospun membranes can be highlighted. Fabric phases, first proposed by Kabir and Furton, consist of the chemical modification by sol-gel reaction of fabrics (cotton, glass fiber) to introduce functional polymers to its surface [15]. On the other hand, electrospun membranes provide the analyst with a wide range of tools since the characteristics of the final product depend on the polymeric precursor(s) and conditions used during the electrospinning [16]. The use of nanoparticles (NPs) as ingredients in these materials makes their application scope even broader. The presence of these NPs usually enhances the sorption capacity by two mechanisms that can be complementary: NPs may introduce new sorption sites in the polymer structure, and may increase the superficial area of the polymer [17].

A NP can be defined as a particle that has at least one dimension in the nanometric range (100 nm is used as a limit by convention) and presents unique properties (not observable in the bulk material) because of its size [18]. There would be a myriad of different NPs if we considered their chemical composition, size, shapes, and potential combinations. From a chemical point of view, NPs can be classified as inorganic or carbon-based. The latter, including fullerenes and nanotubes, have been extensively used in microextraction [19], although the use of single-walled carbon nanohorns (SWCNHs), first described by Ijima et al. in 1991 [20], is limited [21,22]. SWCNHs consist of horn-shaped sheath aggregates of graphene. They usually present lengths in the range of 40–50 nm and an inner diameter from 2 to 5 nm. Their oxidation, to introduce oxygen-containing functional groups to the surface, is easier compared to carbon nanotubes. In solution, SWCNHs are prone to aggregate, forming ordered and stable structures called dahlias [23]. Although the aggregation tendency is common for all carbon nanoparticles, especially when their surface is not chemically modified, these ordered aggregates have been only described for CNHs. This particularity is exploited in the present work.

Oakes et al. proposed in 2013 the electrodeposition of CNHs in different substrates, opening the door for the synthesis of CNHs-coated materials for catalysis, sensing, or energy storage applications [24]. In this article, we propose for the first time the use of SWCNHs-modified paper as a sorptive phase in TFME without the assistance of an anchoring polymer. It has been observed that the dahlia aggregates, obtained in solution, form suprastructures (ordered combinations of single dahlias) when the solvent is evaporated. These suprastructures present a porous conformation that enhances their sorption ability. To make the handling of this sorptive phase easier, it has been coated over conventional paper. It is necessary to indicate that paper acts as a simple support and that suprastructures form over its surface. A small percentage of the suprastructure is embedded in the cellulose fibers, thus improving the mechanical stability of the sorptive phase.

The use of paper as a support opens the door for the development of cheap and disposable units. Meng et al. proposed the use of unmodified paper for the extraction of 8-hydroxy-2′-deoxyguanosine from a urine sample [25], while Saraji and Farajmand have reported the use of modified paper as a sorptive phase [26]. Our research group proposed the direct coating of conventional paper with polymers as a simple strategy to synthesize new sorptive phases [27]. Also, the resulting sorptive phase can be easily cut to the desired shape and length. The potential of paper goes beyond these applications since it can be directly analyzed by MS, in the so-called paper spray MS [28–31], which implies a dramatic simplification of the analytical process. To be used in the paper spray mode, the phases must be conductive and mechanically stable to avoid MS source contamination. As the mechanical stability of this phase in a high-voltage gradient should be previously guaranteed, this initial research will be focused on the evaluation of its sorption ability towards selected antidepressants in urine. To speed up and simplify the analysis, the eluates are directly infused [32,33] in the mass spectrometer, thereby avoiding a previous chromatographic separation.

## 2. Materials and Methods

### 2.1. Reagents

All reagents were of analytical grade or better. Unless otherwise specified, they were purchased from Sigma Chemical Co. (St. Louis, MO, USA, https://www.sigmaaldrich.com/). Stock standard solutions of the antidepressants (mianserin, trimipramine, desipramine, and amitriptyline) were prepared in methanol at a concentration of 1 g/L and stored at 4 °C. Working standard solutions were prepared daily by rigorous dilution of the stocks in ultrapure Milli-Q water. Methanol:acetic acid (95:5) was also used for antidepressants elution after the extraction. Deuterated 5-hydroxyindole-3-acetic acid (5-HIAA-D5) was used as an internal standard for MS measurements. The working concentration of the internal standard was 100 ng/mL.

SWCNHs were purchased from Carbonium S.r.l. (Padua, Italy, www.carbonium.it/). They form stable dahlia-shaped aggregates with an average diameter of 60–80 nm. Individually, the lengths of these SWCNHs are in the range 40–50 nm, and the width in the cylindrical structure varies between 4 and 5 nm. For the synthesis of the sorptive phases, SWCNTs were dispersed in chloroform.

Acetonitrile, acetic acid (Scharlab, Barcelona, Spain, http://www.scharlab.com/), triethylamine, and ultrapure Milli-Q water were employed as components of the chromatographic mobile phase.

Blank urine samples were collected from healthy adult volunteers and stored in polytetrafluoroethylene (PTFE) flasks at −20 °C until analysis. Before the extraction, each sample was 1:1 diluted with ultrapure water and the pH was also adjusted to 10 with sodium hydroxide. The pH is fixed at alkaline conditions to promote the interaction between the basic analytes with the sorptive phase. The interaction between the sorptive phase and the analytes (their chemical structures and the logarithm of the octanol/water partition coefficients are shown in Figure S1) involve general hydrophobic interactions and π–π bonds with the aromatic moieties. The samples are not filtered before their extraction.

### 2.2. Synthesis and Characterization of Sorptive Phases

The synthesis follows a simple workflow. First, 10 mg of SWCNHs are dispersed by manual shaking in 150 μL of chloroform inside an Eppendorf flask. Once dispersed, segments of filter paper (3 × 0.5 cm) are dipped three consecutive times into the dispersion, drying the paper after each dip. The evaporation of the solvent leaves a suprastructure of dahlia aggregates over the paper surface. The resulting materials were characterized by scanning electron microscopy (SEM) using a JEOL JSM 7800 microscope (JEOL, Tokyo, Japan). Micrographs were acquired at the central Service for Research Support (SCAI) of the University of Córdoba.

### 2.3. Microextraction Procedure

A plastic pipette is used to build a simple extraction device. A segment of sorptive phase with an area of 0.25 cm$^2$ is located and mechanically fixed (physically caught in the narrower section) in a 200 μL pipette tip, as indicated in Figure 1. The extraction process comprises several steps. Initially, 1200 μL of the pretreated sample are located in a vial. In the meantime, the sorptive phase is conditioned with 200 μL of methanol and 200 μL of an alkaline aqueous solution (pH = 10), which are aspirated and ejected. Once the sorptive phase is in the best conditions, the tip is immersed in the sample vial, and 80 strokes (aspiration and ejection cycles) are performed to maximize the interaction of the analytes with the membrane. Before the final elution of the analytes using 50 μL of methanol:acetic acid (95:5, *v/v*), the sorptive phase is washed with 200 μL of alkaline aqueous solution (pH = 10). The extracts are analyzed by UPLC-DAD or direct infusion MS for analyte identification and quantification, as indicated in the next section. For direct infusion MS, the internal standard is added to the eluent at a concentration of 100 ng/mL.

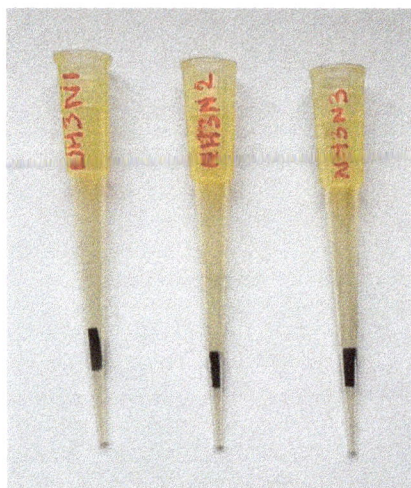

**Figure 1.** Three different pipette tip extraction units containing carbon nanohorn suprastructures coated over conventional paper as a sorptive phase. The phase is mechanically fixed to the narrower section of the tip and in close contact with its inner walls.

*2.4. Instrumental Techniques*

Two instrumental techniques were employed in the development of the present research. The optimization of the extraction procedure and its preliminary analytical evaluation was carried out on a Waters AcquityTM Ultra Performance LC system (Waters Corp., Madrid, Spain) using an Acquity UPLC® BEH C18 column (1.7 μm, 2.1 mm × 100 mm) working at the experimental conditions described in the Supplementary Materials. Direct infusion MS measurements were performed on Agilent 6420 Triple Quadrupole MS with electrospray source using Agilent MassHunter Software (version B.06.00, Santa Clara, CA, USA) for data analyses. The mass spectrometer settings were fixed to improve the SRM signal. The flow rate and the temperature of the drying gas ($N_2$) were 9 L/min and 300 °C, respectively. The nebulizer pressure was 40 psi, and the capillary voltage was kept to 2000 V in positive mode. The analytes and the internal standard were detected by Selected Reaction Monitoring (SRM) transitions, the parameters being specified in Table S1.

## 3. Results and Discussion

*3.1. Synthesis and Characterization of Sorptive Phases*

Paper is an excellent support for the preparation of new microextraction and sensing platforms due to its low price, high porosity, and easy chemical modification. Conventional paper consists of natural cellulose fibers, mechanically compacted, creating a crisscross pattern, as can be observed in Figure 2A. On the other hand, SWCNHs aggregate in solution, forming a stable structure called a dahlia. When the solvent evaporates, these aggregates form suprastructures that consist of self-assembled dahlias. If an SWCNHs dispersion in chloroform is prepared, and a conventional paper is dipped into it, these suprastructures can be created after the solvent evaporation over the paper surface. Dipping, among other approaches, has been proposed for the preparation of coatings in SPME [34]. Figure 2B shows the SEM picture of modified paper where single nanometric dahlias, showing a semispherical shape, are easily identified, while the typical fibrous structure of the unmodified paper is not observed.

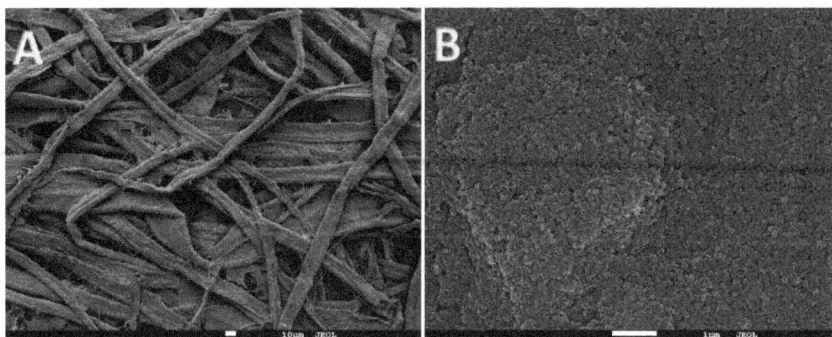

**Figure 2.** SEM micrographs of (**A**) unmodified paper (at 300 magnifications) and (**B**) coated paper (at 13,000× magnification). The presence of the SWNHs suprastructure, which completely covers the cellulose fibers, is observable on the surface.

The thickness of the suprastructure coating can easily be modified by increasing the number of paper dips. Figure 3 shows the superficial SEM pictures of two sorptive phases fabricated using one and three dips, respectively. The thickness of the coating increases from ca. 84 to 190 μm. The thickness is directly related to the extraction capacity, as can be observed for desipramine in Figure S2. The best extraction recoveries are obtained for three dips, which indicates that the sorption is not only superficial; the size of the pores is sufficient for the diffusion of the analytes. Although the extraction is not exhaustive, the results are comparable with those obtained in other microextraction techniques.

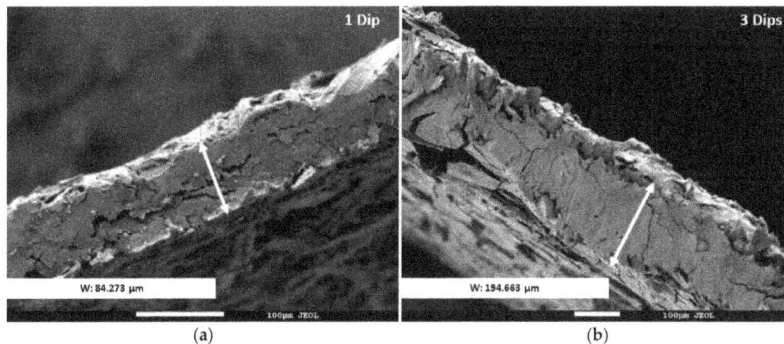

**Figure 3.** SEM micrographs of the side profile of sorptive phases synthesized after (**a**) one dip (at 270 magnifications) and (**b**) three dips (at 140 magnification). The thickness of the coating (considering the different scale of the pictures) increases from ca. 84 to 190 μm.

It is assumed that the mechanical stability is the critical issue of the new phase. However, different tests have demonstrated acceptable stability under working conditions where the sample is passed laterally (it does not flow) through the sorptive phase. On the one hand, after the synthesis, the sorptive phases are cleaned with different solvents, and negligible detachment of the suprastructures is observed. Although the extracts are filtered before their injection in the UPLC, the dahlia sizes (60–80 nm) are smaller than the filter pore size (0.22 μm). A detachment of the nanoparticles would affect the column backpressure (which is very sensitive to the introduction of particles) in UPLC analyses. However, the pressure remained in the normal working range. Finally, sorptive phases can be reused up to 100 times, which indicates that the sorptive phase is not lost during the extraction.

The explanation of this stability can be found on the support. Paper is porous, and the first layer of CNHs aggregates can be occluded on its structure while the subsequent layers can be stabilized by non-covalent (π–π) interactions. This stabilization and the working conditions previously described can be the reason behind this acceptable stability.

*3.2. Extraction Evaluation*

The effect of three critical parameters on the extraction of the analytes was evaluated. The sensitivity and precision enhancement were considered to select the most appropriate conditions. The optimization was done following a one variable at a time approach. Once optimized, the variable is fixed at its optimum value to study the rest of the parameters. Each condition was evaluated in triplicate.

The number of sample strokes (times that the solution is pulled into and ejected from the pipette tip) was evaluated, placing 800 μL of an aqueous standard in a glass vial. As can be observed in Figure 4A, the extraction recovery increases with the number of strokes up to 80 cycles, before decreasing for further values. The number of cycles indicates that the diffusion of the analytes from the solution to the sorptive phases is not promoted. This fact can be ascribed to the geometry of the pipette tip and the location of the phase in one of the inner walls of the tip (see Figure 1), which permits the direct contact of only a fraction of the sample with the sorptive phase. The modification of the extraction unit to enhance this transference is a current line of research.

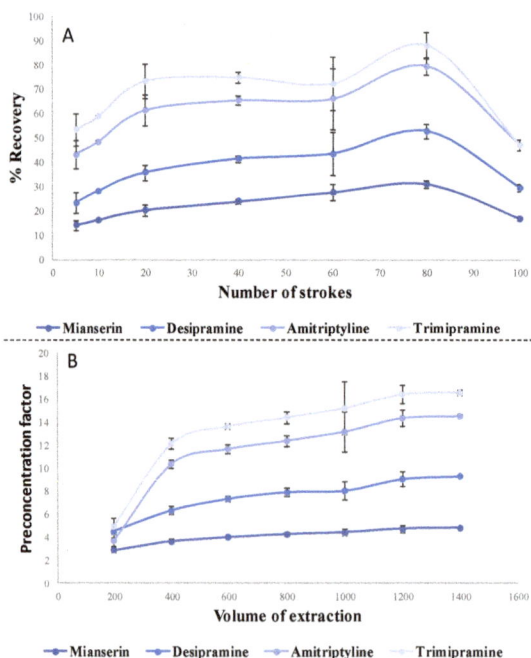

**Figure 4.** Effect of the (**A**) number of strokes and (**B**) sample volume on the extraction recovery of the analyte. The sample volume is defined as the volume of sample placed in the extraction vial. Each condition was evaluated in triplicate.

The sample volume was also considered in depth. This volume was defined as the volume of sample placed in the extraction vial. For the extraction, the pipette tip containing the sorptive phase is immersed in the sample, and 80 strokes are done without taking the tip out of the sample. As can be observed, the peak of the area of the analytes increased with the initial sample volume up to 1200 µL (Figure 4B), which was selected as the optimum value. This volume is compatible with bioanalytical samples like urine or saliva.

Finally, the number of elution strokes was considered. The results (data not shown) indicate that 10 cycles were enough to elute the analytes.

### 3.3. Analytical Evaluation

The combination of the extraction workflow with chromatographic analysis (UPLC-DAD) was initially considered to fully understand the potential of the sample treatment. For this purpose, a calibration curve for each analyte was constructed. Good linearity ($R > 0.995$) was observed in the range 10–1500 µg/L for desipramine, amitriptyline, and mianserin, while trimipramine presented a slightly lower value ($R = 0.989$). The limits of detection, which were calculated using a signal to noise ratio of 3, were in the range of 0.1 µg/L for desipramine, amitriptyline, and mianserin, while trimipramine presented a slightly higher value (0.2 µg/L). The repeatability of the method, expressed as the relative standard deviation (RSD, %), ranged from 3.8% (amitriptyline) to 7.4% (trimipramine). The analysis of a raw urine sample with the method did not show good performance, with relative recoveries (calculated at 200 µg/L) in the 50–60% range, although the 1:1 dilution of the sample fulfilled the 70–130% recovery criterion.

Once the performance of the extraction workflow in combination with UPLC-DAD was evaluated, its direct coupling with MS was also studied. Direct infusion MS allows the reduction of the analysis time, providing good selectivity (working in the SRM mode) and sensitivity (if ion-suppression is negligible). In most cases, ESI-MS requires the use of an internal standard to improve the precision measurement, and in this case, 5-HIAA-D5 was used for this purpose. According to the results obtained with the UPLC-UV combination, in-matrix calibration was selected for the direct infusion approach. Interestingly, the calibration curves obtained for the analytes in the range from 0.1 to 10 µg/L (six different concentration levels, $n = 3$), and prepared in blank urine 1:1 diluted in water, were linear ($R > 0.993$) for almost all the analytes, trimipramine excepted ($R = 0.9$), while the limits of detection were in the range of 10 ng/L for desipramine, amitriptyline, and mianserin. The precision, expressed as RSD values, was evaluated at 0.1 µg/L considering three replicates. The values of the last three analytes were 7.2%, 7.3%, and 9.8%, respectively.

Table 1 summarizes and compares the sensitivity levels provided by different analytical methods [35–40] proposed for the determination of antidepressant drugs in biological fluids. The new approach provided the best results thanks to the use of direct infusion MS as instrumental technique. This combination combines the inherent sensitivity of MS with the higher injection volumes allowed in direct infusion. In fact, the sample volume is limited in chromatographic separation by the resolution factor.

**Table 1.** Comparison of the sensitivity provided by the new method with other counterparts proposed for the determination of antidepressant drugs in biological samples.

| Extraction Procedure [1] | Instrumental Technique [2] | Sample | Linear Range | LOD | Reference |
|---|---|---|---|---|---|
| Micro SPE | LC-UV | Urine | 14–1000 µg/L | 8.6–15.2 µg/L | [35] |
| Hollow fiber drop to drop microextraction | GC-MS | Water Urine Blood | 0.5–50 mg/L | 0.007–0.021 mg/L | [36] |
| Ionic liquid-dispersive liquid-liquid microextraction | LC-MS/MS | Blood | 10–1000 µg/L | 1–2 µg/L | [37] |
| Thin film microextraction | DCBI-MS | Plasma | 5–1000 µg/L | 0.3–1 µg/L | [38] |
| SPME | LC-UV | Urine | 10–400 µg/ | 3–5 µg/ | [39] |
| MEPS | GC-MS | Urine | 0.1–100 µg/L | 0.03–0.05 µg/L | [40] |
| Thin film microextraction | Direct infusion-MS | Urine | 0.1–10 | 10 ng/L | This work |

[1] SPE, solid phase extraction; SPME, solid phase microextraction; MEPS, microextraction in packed sorbent. [2] LC, liquid chromatography; GC, gas chromatography; MS mass spectrometry; DCBI, desorption corona beam ionization; UV, ultraviolet detection.

## 4. Conclusions

This article presents carbon nanohorn suprastructures coated over conventional paper as a sorptive phase in thin film microextraction. To make a critical and complete study of the new phase, a SWOT (Strengths, Weaknesses, Opportunities, and Threats) analysis has been done. This study, which is schematically presented in Figure 5, is focused on the new sorptive phase rather than the analytical application.

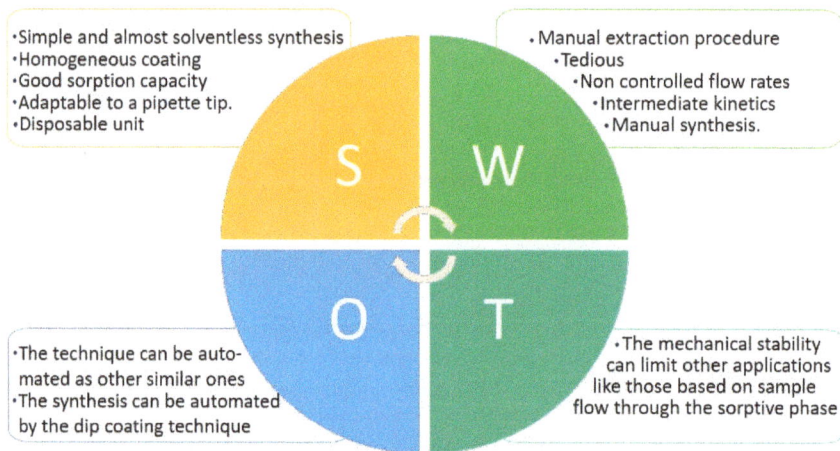

**Figure 5.** SWOT (Strengths, Weaknesses, Opportunities, and Threats) analysis of the new sorptive phase.

The synthesis is simple as it only requires dipping the paper into an organic dispersion containing the nanostructures. The evaporation of the solvent leaves a continuous and homogeneous layer of aggregated dahlias over the paper surface, which can interact with the target analytes. The volume of solvent is very low, and the synthesis can be considered almost solventless. The as-prepared sorptive phases, which have dimensions of $3 \times 0.5$ cm, are finally adapted to conventional pipette tips, which act as simple extraction devices. The volume of the tips and their disposable nature make this approach attractive in the bioanalytical context. Several variables, including the number of dips, strokes, or sample volume, have been considered in detail to fully understand the potential of the sorptive phase. The combination of the microextraction technique and direct infusion MS allows the rapid detection and determination of three antidepressants (desipramine, amitriptyline, and mianserin) in urine samples with limits of detection in the ng/L range. Considering the chemical characteristics of the SWNHs, the new sorptive phase has the potential to interact with a great variety of compounds, especially those containing aromatic rings on their structures.

In this first approach a manual extraction is performed; therefore, the procedure is tedious and the sample flow rate cannot be efficiently controlled. As has been demonstrated, many strokes are required for the extraction, which indicates intermediate kinetics (the velocity of the extraction depends on the tip geometry as only a fraction of the aspirated volume meets the sorptive phase). In the same way, the synthesis (dip and evaporation) is done manually.

The potential automation of pipette tip extraction [41–43] or the use of static extraction procedures (where several samples can be extracted at the same time) may be a solution. Also, the synthesis can be automated by dip coating technology.

The stability of the coating is the main limitation for flow through applications, as the CNHs superstructure is not covalently bonded to the paper substrate. Also, assuring their stability in the paper-spray MS approach will be an exciting challenge in the near future.

**Supplementary Materials:** The following are available online. Figure S1: Chemical structure of the four antidepressants drugs studied in this work. The logarithm of octanol/water partition coefficients (log P) for all the analytes at the working pH are also shown (source www.chemspider.com), Figure S2: Effect of the number of dips into the extraction recovery of the target analytes, Table S1: Selected reaction monitoring parameters for the MS analyses.

**Author Contributions:** J.R.-G., B.F.-C. and R.L. designed the extraction experiments; M.T.G.-V designed the direct infusion MS experiments; J.R.-G. performed the experiments; S.C. and R.L. supervised the project; all the authors contributed to the paper writing.

**Funding:** Financial support from the Spanish Ministry of Economy and Competitiveness (CTQ2017-83175R) is gratefully acknowledged.

**Acknowledgments:** J.R.-G., B.F.-C., and M.T.G.-V. express their gratitude for the predoctoral grants (FPU13/03549, FPU13/03896, and BES-2015-071421) from the Spanish Ministry of Education. The authors would like to thank the Central Service for Research Support (SCAI) of the University of Córdoba for the service provided to obtain the micrographs. Article processing charge was sponsored by MDPI.

**Conflicts of Interest:** The authors declare no conflict of interest.

# References

1. Arthur, C.L.; Pawliszyn, J. Solid phase microextraction with thermal desorption using fused silica optical fibers. *Anal. Chem.* **1990**, *62*, 2145–2148. [CrossRef]
2. Souza-Silva, É.A.; Gionfriddo, E.; Shirey, R.; Sidisky, L.; Pawliszyn, J. Methodical evaluation and improvement of matrix compatible PDMS-overcoated coating for direct immersion solid phase microextraction gas chromatography (DI-SPME-GC)-based applications. *Anal. Chim. Acta* **2016**, *920*, 54–62. [CrossRef] [PubMed]
3. Gionfriddo, E.; Boyacl, E.; Pawliszyn, J. New Generation of Solid-Phase Microextraction Coatings for Complementary Separation Approaches: A Step toward Comprehensive Metabolomics and Multiresidue Analyses in Complex Matrices. *Anal. Chem.* **2017**, *89*, 4046–4054. [CrossRef] [PubMed]
4. Gómez-Ríos, G.A.; Liu, C.; Tascon, M.; Reyes-Garcés, N.; Arnold, D.W.; Covey, T.R.; Pawliszyn, J. Open Port Probe Sampling Interface for the Direct Coupling of Biocompatible Solid-Phase Microextraction to Atmospheric Pressure Ionization Mass Spectrometry. *Anal. Chem.* **2017**, *89*, 3805–3809. [CrossRef] [PubMed]
5. Gómez-Ríos, G.A.; Vasiljevic, T.; Gionfriddo, E.; Yu, M.; Pawliszyn, J. Towards on-site analysis of complex matrices by solid-phase microextraction-transmission mode coupled to a portable mass spectrometer via direct analysis in real time. *Analyst* **2017**, *142*, 2928–2935. [CrossRef] [PubMed]
6. Bruheim, I.; Liu, X.; Pawliszyn, J. Thin-film microextraction. *Anal. Chem.* **2003**, *75*, 1002–1010. [CrossRef] [PubMed]
7. Jiang, R.; Pawliszyn, J. Thin-film microextraction offers another geometry for solid-phase microextraction. *TrAC Trends Anal. Chem.* **2012**, *39*, 245–253. [CrossRef]
8. Riazi Kermani, F.; Pawliszyn, J. Sorbent coated glass wool fabric as a thin film microextraction device. *Anal. Chem.* **2012**, *84*, 8990–8995. [CrossRef] [PubMed]
9. Mirnaghi, F.S.; Pawliszyn, J. Development of coatings for automated 96-blade solid phase microextraction-liquid chromatography-tandem mass spectrometry system, capable of extracting a wide polarity range of analytes from biological fluids. *J. Chromatogr. A* **2012**, *1261*, 91–98. [CrossRef] [PubMed]
10. Vuckovic, D.; Cudjoe, E.; Musteata, F.M.; Pawliszyn, J. Automated solid-phase microextraction and thin-film microextraction for high-throughput analysis of biological fluids and ligand-receptor binding studies. *Nat. Protoc.* **2010**, *5*, 140–161. [CrossRef] [PubMed]
11. Karimi, S.; Talebpour, Z.; Adib, N. Sorptive thin film microextraction followed by direct solid state spectrofluorimetry: A simple, rapid and sensitive method for determination of carvedilol in human plasma. *Anal. Chim. Acta* **2016**, *924*, 45–52. [CrossRef] [PubMed]
12. Richter, P.; Leiva, C.; Choque, C.; Giordano, A.; Sepúlveda, B. Rotating-disk sorptive extraction of nonylphenol from water samples. *J. Chromatogr. A* **2009**, *1216*, 8598–8602. [CrossRef] [PubMed]
13. Alcudia-León, M.C.; Lucena, R.; Cárdenas, S.; Valcárcel, M. Stir membrane extraction: A useful approach for liquid sample pretreatment. *Anal. Chem.* **2009**, *81*, 8957–8961. [CrossRef] [PubMed]
14. Roldán-Pijuán, M.; Lucena, R.; Cárdenas, S.; Valcárcel, M.; Kabir, A.; Furton, K.G. Stir fabric phase sorptive extraction for the determination of triazine herbicides in environmental waters by liquid chromatography. *J. Chromatogr. A* **2015**, *1376*, 35–45. [CrossRef] [PubMed]

15. Racamonde, I.; Rodil, R.; Quintana, J.B.; Sieira, B.J.; Kabir, A.; Furton, K.G.; Cela, R. Fabric phase sorptive extraction: A new sorptive microextraction technique for the determination of non-steroidal anti-inflammatory drugs from environmental water samples. *Anal. Chim. Acta* **2015**, *865*, 22–30. [CrossRef] [PubMed]

16. Reyes-Gallardo, E.M.; Lucena, R.; Cárdenas, S. Electrospun nanofibers as sorptive phases in microextraction. *TrAC Trends Anal. Chem.* **2016**, *84*, 3–11 [CrossRef]

17. Bagheri, H.; Roostaie, A. Roles of inorganic oxide nanoparticles on extraction efficiency of electrospun polyethylene terephthalate nanocomposite as an unbreakable fiber coating. *J. Chromatogr. A* **2015**, *1375*, 8–16. [CrossRef] [PubMed]

18. Auffan, M.; Rose, J.; Bottero, J.Y.; Lowry, G.V.; Jolivet, J.P.; Wiesner, M.R. Towards a definition of inorganic nanoparticles from an environmental, health and safety perspective. *Nat. Nanotechnol.* **2009**, *4*, 634–641. [CrossRef] [PubMed]

19. Valcárcel, M.; Cárdenas, S.; Simonet, B.M.; Moliner-Martínez, Y.; Lucena, R. Carbon nanostructures as sorbent materials in analytical processes. *TrAC Trends Anal. Chem.* **2008**, *27*, 34–43. [CrossRef]

20. Iijima, S. Helical microtubules of graphitic carbon. *Nature* **1991**, *354*, 56–58. [CrossRef]

21. Jiménez-Soto, J.M.; Cárdenas, S.; Valcárcel, M. Oxidized single-walled carbon nanohorns as sorbent for porous hollow fiber direct immersion solid-phase microextraction for the determination of triazines in waters. *Anal. Bioanal. Chem.* **2013**, *405*, 2661–2669. [CrossRef] [PubMed]

22. Fresco-Cala, B.; Cárdenas, S.; Herrero-Martínez, J.M. Preparation of porous methacrylate monoliths with oxidized single-walled carbon nanohorns for the extraction of nonsteroidal anti-inflammatory drugs from urine samples. *Microchim. Acta* **2017**, *184*, 1863–1871. [CrossRef]

23. Iijima, S.; Yudasaka, M.; Yamada, R.; Bandow, S.; Suenaga, K.; Kokai, F.; Takahashi, K. Nano-aggregates of single-walled graphitic carbon nano-horns. *Chem. Phys. Lett.* **1999**, *309*, 165–170. [CrossRef]

24. Oakes, L.; Westover, A.; Mahjouri-Samani, M.; Chatterjee, S.; Puretzky, A.A.; Rouleau, C.; Geohegan, D.B.; Pint, C.L. Uniform, homogenous coatings of carbon nanohorns on arbitrary substrates from common solvents. *ACS Appl. Mater. Interfaces* **2013**, *5*, 13153–13160. [CrossRef] [PubMed]

25. Meng, X.; Liu, Q.; Ding, Y. Paper-based solid-phase microextraction for analysis of 8-hydroxy-2′-deoxyguanosine in urine sample by CE-LIF. *Electrophoresis* **2017**, *38*, 494–500. [CrossRef] [PubMed]

26. Saraji, M.; Farajmand, B. Chemically modified cellulose paper as a thin film microextraction phase. *J. Chromatogr. A* **2013**, *1314*, 24–30. [CrossRef] [PubMed]

27. Ríos-Gómez, J.; Lucena, R.; Cárdenas, S. Paper supported polystyrene membranes for thin film microextraction. *Microchem. J.* **2017**, *133*, 90–95. [CrossRef]

28. Damon, D.E.; Davis, K.M.; Moreira, C.R.; Capone, P.; Cruttenden, R.; Badu-Tawiah, A.K. Direct Biofluid Analysis Using Hydrophobic Paper Spray Mass Spectrometry. *Anal. Chem.* **2016**, *88*, 1878–1884. [CrossRef] [PubMed]

29. Mendes, T.P.P.; Pereira, I.; Ferreira, M.R.; Chaves, A.R.; Vaz, B.G. Molecularly imprinted polymer-coated paper as a substrate for highly sensitive analysis using paper spray mass spectrometry: Quantification of metabolites in urine. *Anal. Methods* **2017**, *9*, 6117–6123. [CrossRef]

30. Wang, T.; Zheng, Y.; Wang, X.; Austin, D.E.; Zhang, Z. Sub-ppt Mass Spectrometric Detection of Therapeutic Drugs in Complex Biological Matrixes Using Polystyrene-Microsphere-Coated Paper Spray. *Anal. Chem.* **2017**, *89*, 7988–7995. [CrossRef] [PubMed]

31. Pereira, I.; Rodrigues, M.F.; Chaves, A.R.; Vaz, B.G. Molecularly imprinted polymer (MIP) membrane assisted direct spray ionization mass spectrometry for agrochemicals screening in foodstuffs. *Talanta* **2018**, *178*, 507–514. [CrossRef] [PubMed]

32. Castrillo, J.I.; Hayes, A.; Mohammed, S.; Gaskell, S.J.; Oliver, S.G. An optimized protocol for metabolome analysis in yeast using direct infusion electrospray mass spectrometry. *Phytochemistry* **2003**, *62*, 929–937. [CrossRef]

33. González-Domínguez, R.; García-Barrera, T.; Gómez-Ariza, J.L. Using direct infusion mass spectrometry for serum metabolomics in Alzheimer's disease. *Anal. Bioanal. Chem.* **2014**, *406*, 7137–7148. [CrossRef] [PubMed]

34. Aziz-Zanjani, M.O.; Mehdinia, A. A review on procedures for the preparation of coatings for solid phase microextraction. *Microchim. Acta* **2014**, *181*, 1169–1190. [CrossRef]

*Molecules* **2018**, *23*, 1252

35. Fresco-Cala, B.; Mompó-Roselló, Ó.; Simó-Alfonso, E.F.; Cárdenas, S.; Herrero-Martínez, J.M. Carbon nanotube-modified monolithic polymethacrylate pipette tips for (micro) solid-phase extraction of antidepressants from urine samples. *Microchim. Acta* **2018**, *185*, 127. [CrossRef] [PubMed]
36. Jagtap, P.K.; Tapadia, K. Pharmacokinetic determination and analysis of nortriptyline based on GC–MS coupled with hollow-fiber drop-to-drop solvent microextraction technique. *Bioanalysis* **2018**, *10*, 143–152. [CrossRef] [PubMed]
37. De Boeck, M.; Dubrulle, L.; Dehaen, W.; Tytgat, J.; Cuypers, E. Fast and easy extraction of antidepressants from whole blood using ionic liquids as extraction solvent. *Talanta* **2018**, *180*, 292–299. [CrossRef] [PubMed]
38. Chen, D.; Hu, Y.-N.; Hussain, D.; Zhu, G.-T.; Huang, Y.-Q.; Feng, Y.-Q. Electrospun fibrous thin film microextraction coupled with desorption corona beam ionization-mass spectrometry for rapid analysis of antidepressants in human plasma. *Talanta* **2016**, *152*, 188–195. [CrossRef] [PubMed]
39. Rajabi, A.A.; Yamini, Y.; Faraji, M.; Seidi, S. Solid-phase microextraction based on cetyltrimethylammonium bromide-coated magnetic nanoparticles for determination of antidepressants from biological fluids. *Med. Chem. Res.* **2013**, *22*, 1570–1577. [CrossRef]
40. Bagheri, H.; Banihashemi, S.; Zandian, F.K. Microextraction of antidepressant drugs into syringes packed with a nanocomposite consisting of polydopamine, silver nanoparticles and polypyrrole. *Microchim. Acta* **2016**, *183*, 195–202. [CrossRef]
41. Guan, H.; Brewer, W.E.; Garris, S.T.; Morgan, S.L. Disposable pipette extraction for the analysis of pesticides in fruit and vegetables using gas chromatography/mass spectrometry. *J. Chromatogr. A* **2010**, *1217*, 1867–1874. [CrossRef] [PubMed]
42. Brewer, W.E. Disposable pipette extraction. U.S. Patent US6566145B2, 20 May 2003.
43. Brewer, W.E. Dispersive pipette extraction tip and methods for use. U.S. Patent US9733169B2, 15 Augus 2017.

**Sample Availability:** Samples of the membranes are available from the authors.

# molecules

MDPI

*Article*

# In-Syringe Micro Solid-Phase Extraction Method for the Separation and Preconcentration of Parabens in Environmental Water Samples

Geaneth Pertunia Mashile, Anele Mpupa and Philiswa Nosizo Nomngongo *

Department of Applied Chemistry, University of Johannesburg, Doornfontein Campus, P.O. Box 17011, Johannesburg 2028, South Africa; petmashile2009@hotmail.com (G.P.M.); anelempupa@yahoo.com (A.M.)
* Correspondence: pnnomngongo@uj.ac.za or nomngongo@yahoo.com; Tel.: +27-11-559-6187

Academic Editor: Victoria F. Samanidou
Received: 22 May 2018; Accepted: 13 June 2018; Published: 14 June 2018

**Abstract:** In this study, a simple, rapid and effective in-syringe micro-solid phase extraction (MSPE) method was developed for the separation and preconcetration of parabens (methyl, ethyl, propyl and butyl paraben) in environmental water samples. The parabens were determined and quantified using high performance liquid chromatography and a photo diode array detector (HPLC-PDA). Chitosan-coated activated carbon (CAC) was used as the sorbent in the in-syringe MSPE device. A response surface methodology based on central composite design was used for the optimization of factors (eluent solvent type, eluent volume, number of elution cycles, sample volume, sample pH) affecting the extraction efficiency of the preconcentration procedure. The adsorbent used displayed excellent absorption performance and the adsorption capacity ranged from 227–256 mg g$^{-1}$. Under the optimal conditions the dynamic linear ranges for the parabens were between 0.04 and 380 µg L$^{-1}$. The limits of detection and quantification ranged from 6–15 ng L$^{-1}$ and 20–50 ng L$^{-1}$, respectively. The intraday (repeatability) and interday (reproducibility) precisions expressed as relative standard deviations (%RSD) were below 5%. Furthermore, the in-syringe MSPE/HPLC procedure was validated using spiked wastewater and tap water samples and the recoveries ranged between from 96.7 to 107%. In conclusion, CAC based in-syringe MSPE method demonstrated great potential for preconcentration of parabens in complex environmental water.

**Keywords:** in-syringe micro solid-phase extraction; personal care products; response surface methodology; parabens; wastewater

---

## 1. Introduction

Over the years living standards have improved and so has the production and use of cosmetics. Cosmetics are defined as any substance or mixture intended for use on the external parts of the human body, teeth and mucous membrane as defined by the European Union 1223/2009 (Article 2,1.Aa) [1]. This mainly includes personal care products (lotions, soup, perfume, deodorant), beauty products (nail polish, mascara, lip stick, etc.), hair care products (shampoo, hair dyes and gel, among others), oral products (mouthwash, toothpaste) [2]. The compounds or chemical substances within the cosmetics are connected to their intended use, that is, antioxidants, preservatives, fragrances, pigment, antimicrobial and UV-filters [3]. One of the most commonly used compounds in cosmetics, pharmaceuticals and personal care products (PPCPs) are parabens. These compounds are used primarily for their antimicrobial and antibacterial properties [4]. In cosmetics and personal care products the most used parabens are methyl, ethyl, propyl as well as butyl paraben which are often times used in combination with other preservatives [5]. They are preferred over other alternatives due to their low cost, broad spectrum of activity, thermal stability and applicability over a wide pH range [6].

Traditionally, parabens have been considered as low toxicity compounds, however more recent studies have found that certain parabens exhibit endocrine disrupting effects, which in turn can lead to a potential increase in breast cancer incidence [7] or the development of malignant melanomas [8], amongst other effects. This is compounded by the ability of parabens to be absorbed into the human skin due to dermal exposure to products which contain these compounds [4]. Moreover, parabens contain phenolic hydroxyl groups, which are capable of producing chlorinated degradation by-products when in contact with chlorinated water as in tap and swimming pool water [9]. The chlorinated by-products which originate from personal care products are said to be more toxic to aquatic organisms that their corresponding parent compounds [10]. The wide use of products containing such compounds has contributed to the direct introduction of parabens into the aquatic environment via the domestic and industrial wastewater route [11]. As a result this has prompted the development of various analytical methods for the determination of parabens in environmental samples (such as water, sewage influents and effluents) [12], soil [13,14], among others.

Over the years various analytical techniques for the determination of parabens in different matrices have been developed, including gas-chromatography (GC) [9,15,16], high performance liquid chromatography (HPLC) [13,17], capillary zone electrophoresis (CZE) [4] and ultra performance liquid chromatography [18,19] amongst others. HPLC methods are the most widely used for the analysis and determination of parabens since techniques such as GC require prior derivatization of the compounds [20].

These analytical techniques are often combined with pretreatment procedures to eliminate non-polar matrix components and also improve the selectivity, reliability and accuracy of the analysis [21,22]. As a result, various extraction and preconcentration techniques have to be used for analysis of parabens such as magnetic solid phase extraction (MSPE) [23], solid phase extraction (SPE) [19]; dispersive liquid liquid microextraction (DLLME) [24], stir-bar sorptive extraction (SBSE) [25] and solid phase microextraction (SPME) [26], amongst others.

Some conventional pretreatment procedures like LLE are scarcely used in the determination of parabens due to their tediousness and high consumption of solvents, so even when automated, the column length limits the ability to separate the different parabens [27]. Thus, conventional SPE has been used as a better alternative to solve these issues, which has resulted in an increase in the application of SPE for the determination of parabens [28]. SPE provides advantages such as preconcentration of trace analytes from larger sample volumes, reduction and elimination of matrix interferences, as well as less possibility of cross-contamination, simplicity, effectiveness of extraction, sensibility and high selectivity [20,22]. Moreover it can easily be automated with a wide variety of available sorbents which have been developed over the years [28]. However, SPE also presents some drawbacks such as limitations in treatment of large sample volumes, its time consuming nature, need for pre-treatment of sorbents prior to extraction, obstruction and clogging of cartridges and relatively low extraction efficiencies. In addition, aggregation [29] of particles may occur reducing the active surface area [30]. SPE cartridges used are usually made from plastics, which can absorb the analytes and hence increase the interference in the analysis [31]. Therefore, to eliminate the drawbacks of conventional SPE, sample pretreatment methods like solid phase microextraction (SMPE) were developed [32]. SPME was proposed as the better alternative as it allows the combination of sampling and sample extraction in a single step, which also eliminates the use of organic solvents for analysis of parabens. It is also a technique that is effective due to its advantages of high enrichment factor, sample operation and solvent-free microextraction [33].

Therefore, the aim of this work was to develop a quick, low solvent consumption, simple and easy to use analytical method based on in-syringe micro-solid phase extraction (MSPE) for the simultaneous preconcentration and determination of four common parabens in environmental matrices. A chitosan-coated activated carbon (CAC) composite was applied as a sorbent in SPME for the preconcentration of parabens in environmental water samples. Activated carbon was chosen due to its high adsorption capacity and affinity for organic compounds/pollutants [34]. This combination improves the performance and serves as a proper support for chitosan as an adsorbent by improving its mechanical and chemical properties. Moreover an added advantage is the lesser quantity of chitosan needed to create a new adsorbent without disrupting the adsorption capacity of the formed adsorbent significantly [35]. To the best of our knowledge, there are limited or no reports on the simultaneous extraction of parabens using a chitosan-coated activated carbon-based in-syringe MSPE method. The efficiency of the preconcentration method for extraction of parabens is due to the unique properties of the adsorbent. A response surface methodology (RSM)-based central composite design (CCD) was used to determine the optimum experimental conditions.

## 2. Results and Discussion

### 2.1. Characterization of Chitosan-Coated Activated Carbon (CAC)

The BET surface area of chitosan-coated activated carbon (CAC) is higher than that of activated carbon. An improvement on the properties of activated carbon was possible thanks to the presence of chitosan acting as a support for coating AC. A result of coating the chitosan is an increase in the surface area of the activated carbon while the pore volume and pore size are decreased due to the fact the chitosan can block the pores on the activated carbon, as seen in Table 1.

**Table 1.** Characteristics of adsorbent.

| Parameter | Activated Carbon | Chitosan Coated Activated Carbon |
|---|---|---|
| BET surface area (m$^2$/g) | 1075.45 | 1181 |
| Pore Volume (cm$^3$/g) | 0.7553 | 0.733 |
| Pore size (nm) | 4.839 | 4.545 |

### 2.2. Optimization of in-Syringe SPME

The optimization of factors affecting the experimental preconcentration procedure was carried out by means of an experimental design approach, whereby the effects of important variables such as mass of adsorbent, sample pH and eluent volume were determined using response surface methodology (RSM) based on a central composite design (CCD). The analysis of variance (ANOVA) represented in terms of Pareto chart was used to explore the significance of the effects on the in-syringe MSPE procedure (Figure 1).

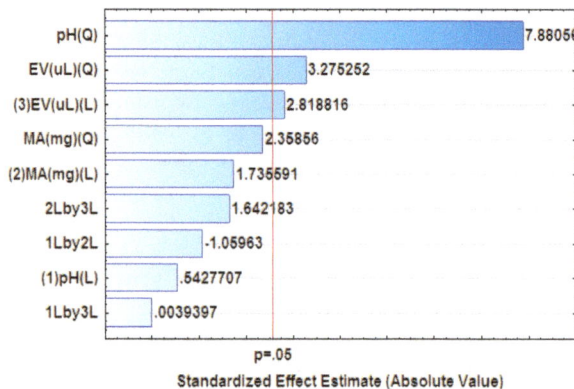

**Figure 1.** Pareto chart of standardized effects for variables in the preconcentration for parabens: The "Q" and "L" in brackets indicate whether the effects are quadratic or linear, respectively. The "2Lby3L" indicates the linear interactions between MA and EV, the "1Lby2L" referrers to the interactions between pH and MA and "1Lby3L" is for the the pH and EV interactions.

According to Figure 1 the effect of pH and mass of adsorbent were insignificant at the 95% confidence level, contrary to the eluent volume used for the preconcentration of parabens. The sample pH and all interactions were not significance at a 95% confidence level for the preconcentration of parabens. RSM was used to establish an extrapolative model which denotes changes in the response, depending on the contributing factors. The quadratic equation for the model illustrated the dependence of the analytical response (%recovery) with respect to the evaluated variables [36].

The 3D response surface plots showing the analytical response against individual factors are shown in Figure 2, which illustrates the interaction of the pH with the mass of adsorbent and eluent volume as well as the interaction of eluent volume with mass of adsorbent, respectively. As seen an enhanced analytical response (%recovery) at pH values between 5 and 8 was observed. This demonstrated that the extraction and pre-concentration of parabens depended on the sample pH. In addition, it was observed that maximum recoveries were achieved at eluent volumes above 500 µL. This is because volumes below 500 µL were not enough to elute all the analytes from the adsorbent. The percentage recovery dependency as a function of the interaction between mass of adsorbent and pH revealed that the analytical response increased with increasing mass of adsorbent. This was probably due to the high surface area and small particle size of the chitosan-coated activated carbon.

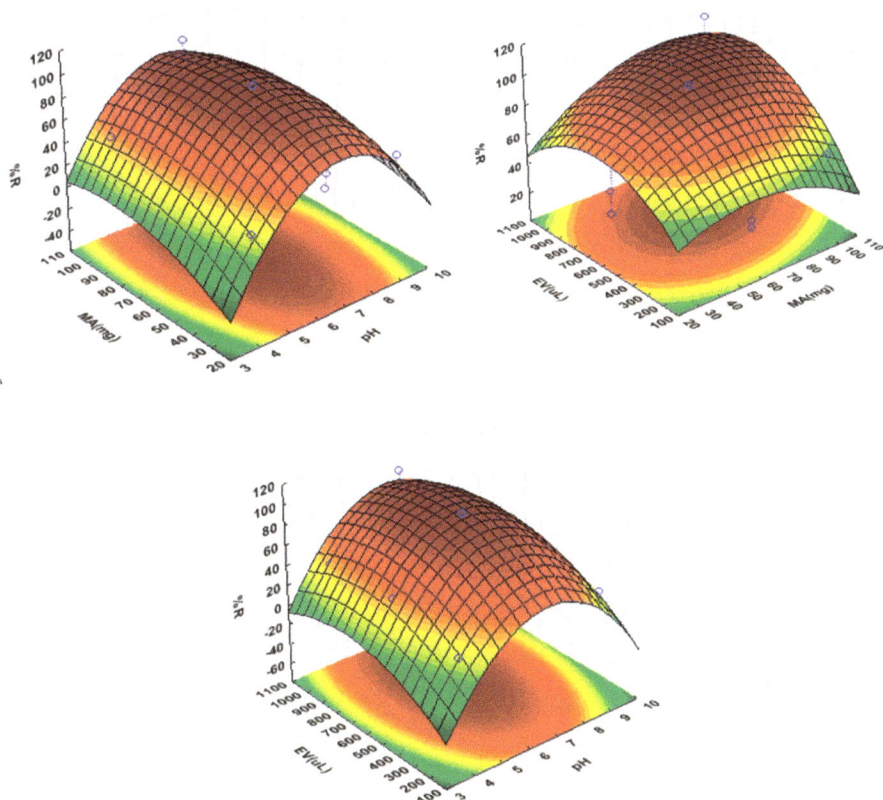

**Figure 2.** Response surface obtained for parabens after extraction and preconcentration by in-syringe MSPE.

In addition to the 3D response surface plots, the estimation of the optimum conditions was further carried out using the profile for predicted values and desirability. As seen in Figure 3 the desirability of 1.0 was assigned for maximum recoveries (105%), 0.0 for minimum (26.5%) and 0.5 for middle (65.8%). Figure 3 also shows the individual desirability scores for preconcentration of parabens (left hand side (bottom) and the, the desirability value of 1.0 was chosen as the target value to be used to obtain optimum conditions and the percent recovery obtained from plots for each parameter in the model is shown at the top left hand side. According to Khodadoust et al. [37], the figures on the top left hand side illustrate the changes in the level of each individual factor and its analytical response as well as its overall desirability. Therefore, based on the calculations from the 3D plots and desirability score of 1.0, the percentage recovery of parameters was optimized at 101.5 and the optimum conditions were mass of adsorbent 81 mg, eluent volume 800 and sample pH 6.5.

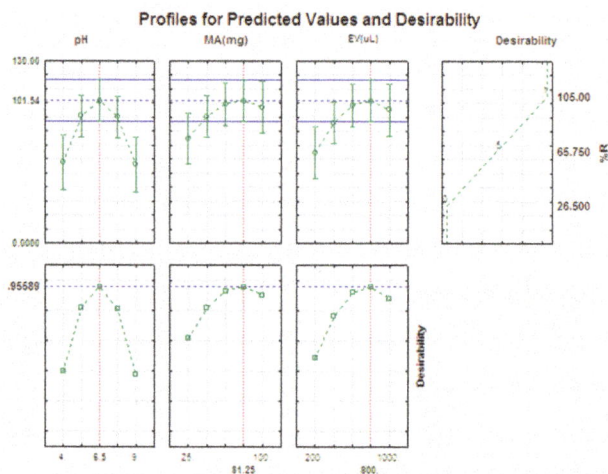

**Figure 3.** Profiles for predicated values and desirability function for pre-concentration of parabens.

### 2.3. Adsorption Capacity of CAC

Under optimum conditions, the adsorption capacities of the CAC toward the parabens were investigated. To evaluate the adsorption capacity, 100 mg $L^{-1}$ of paraben mixture solution was sonicated for 100 min at room temperature and the concentration at equilibrium were determined using HPLC-PDA. The adsorption capacities for each analyte are presented in Table 2.

**Table 2.** Adsorption capacity (mg/g) of chitosan coated activated carbon.

| Paraben | Adsoption Capacity (mg/g) |
|---------|---------------------------|
| Methyl  | 227 |
| Ethyl   | 236 |
| Propyl  | 256 |
| Butyl   | 241 |

## 3. Analytical Figures of Merit

Under optimal conditions the analytical performance of the proposed method was investigated in terms of precision (intra- and interday), linearity, limits of detection (LOD) and limits of quantification (LOQ) (Table 3). Seven point (triplicate) calibration curves for the parabens were constructed by plotting the peak area of the signal acquired using HPLC/PDA as a function of concentrations of parabens.

**Table 3.** Analytical figures of merit of the proposed in-syringe MSPE/HPLC method for the preconcentration and determination of parabens.

| Paraben | Linearity ($\mu$g L$^{-1}$) | Correlation Coefficient ($R^2$) | LOD (ng L$^{-1}$) | LOQ ($\mu$g L$^{-1}$) | Precision (%RSD) | |
|---|---|---|---|---|---|---|
| | | | | | Intraday | Interday |
| Methyl | 0.05–375 | 0.9991 | 12 | 40 | 2.1 | 3.3 |
| Ethyl | 0.04–350 | 0.9989 | 10 | 33 | 1.8 | 2.5 |
| Propyl | 0.04–380 | 0.9995 | 6 | 20 | 1.6 | 3.9 |
| Butyl | 0.06–380 | 0.9992 | 15 | 50 | 1.5 | 2.0 |

**Table 4.** Trueness experiment using three level concentrations of each analyte.

| Paraben | Added Value (ng L$^{-1}$), $n = 5$ | Measured Value (ng L$^{-1}$), $n = 5$ | %RSD | Trueness | |
|---|---|---|---|---|---|
| | | | | %Recovery | %Relative Bias |
| Methyl | 10 | 9.87 ± 0.17 | 1.7 | 98.7 | −1.3 |
| | 30 | 30.5 ± 1.3 | 4.3 | 102 | 1.7 |
| | 50 | 49.8 ± 1.8 | 3.6 | 99.6 | −0.4 |
| Ethyl | 10 | 9.92 ± 0.23 | 2.3 | 99.2 | −0.8 |
| | 30 | 29.8 ± 0.5 | 1.7 | 99.3 | −0.7 |
| | 50 | 48.9 ± 1.5 | 3.1 | 97.8 | −2.2 |
| Propyl | 10 | 9.91 ± 0.12 | 1.2 | 99.1 | −0.9 |
| | 30 | 29.2 ± 0.6 | 2.1 | 97.3 | −2.7 |
| | 50 | 50.3 ± 1.3 | 2.6 | 101 | 0.6 |
| Butyl | 10 | 9.76 ± 0.16 | 1.6 | 97.6 | −2.4 |
| | 30 | 29.7 ± 0.9 | 3.0 | 99.0 | −1.0 |
| | 50 | 50.1 ± 1.2 | 2.4 | 100 | 0.2 |

*Molecules* **2018**, 23, 1450

Calibration curve linearity was observed in the range of 0.05–375 µg L$^{-1}$ for methylP, 0.04–350 µg L$^{-1}$ for ethylP, 0.04–380 µg L$^{-1}$ for propylP and butylP 0.06–380 µg L$^{-1}$. The correlation coefficient ($R^2$) ranged from 0.9989 to 0.9995 (Table 4). The LODs, LOQs, enrichment factor and pre-concentration factor ranged 6–15 ng L$^{-1}$, 20–50 ng L$^{-1}$, 100 and 150–175, respectively.

The intraday precision (repeatability) was evaluated by analyzing fifteen successive replicates 50 µg L$^{-1}$. The precision expressed in the form of relative standard deviation (%RSD) ranged from 1.5 to 2.1%. In addition, the interday precision (reproducibility of in-syringe device, triplicates, five working days) were between 2.0% and 3.9%.

Due to the absence of certified reference materials (CRM) for parabens in water samples, spike recovery experiments were performed for assessing the trueness of the developed method. The trueness validation procedure was carried out by spiking tap water at three concentration levels, that is low (10 ng L$^{-1}$), middle (30 ng L$^{-1}$), and high (50 ng L$^{-1}$). Samples were analyzed in triplicate over a period of five days and the trueness of the method was assessed using relative bias and recovery (Table 4). As seen from Table 4, the relative bias varied between −2.7% and 1.7% while the recoveries were greater than 95% for all the studied analytes. In addition, the inter-day precision was less than 5% and these demonstrated the trueness of the developed method.

The analytical performance of the proposed in-syringe MSPE method for preconcentration and determination of parabens was compared with other different analytical procedures reported in the literature (Table 5). Thus, it can be observed that the proposed method has better analytical figures of merit compared to those reported by [19,38–44]. In addition, the analytical performances were comparable to those reported by [41] and lower than those reported by [42]. The results obtained in this work show that the in-syringe SPME-HPLC-DAD is efficient and a simple technique. Its optimum performances can be attributed to the attractive properties of the adsorbent. Thus, the method proved to have higher adsorption capacity, simple, cost effective and sensitive and efficient for the preconcentration and determination of parabens in water samples.

**Table 5.** Comparison of the proposed microextraction method with literature for determination of parabans.

| Parabens | Matrices | Analytical Method | LOD (ng L$^{-1}$) | References |
|---|---|---|---|---|
| Methyl, ethyl, propyl, isopropyl, butyl, isobutyl, benzyl | Wastewater, River water | SPE/HPLC-CL | 0.08–0.44 | [42] |
| Methyl, ethyl, propyl | Cosmetics, beverages, water | HPLC-UV/ SUPRAS | 70–500 | [43] |
| Methyl, propyl, butyl, benzyl and benzophenone-4 | Wastewater (Influent and effluent) | SPE-HPLC-MS/MS | 0.8–4.8 | |
| Methyl, ethyl, propyl, isopropyl, butyl, benzyl | Wastewater | SPE/UPLC-MS/MS | 221–21,423 | [19] |
| Methyl, ethyl, propyl, butyl | Wastewater, river water, swimming pool water | DLLME/GC-MS/MS | 3.90–27.5 | [40] |
| Methyl, ethyl, propyl | Toothpaste, mouth rinse, shampoo, tap water, river water | DLLME/HPLC-UV | 5–20 | [41] |
| Methyl, propyl | Underground water | HF-MMLLE/HPLC-DAD | 500–4600 | [39] |
| Ethyl, propyl, isobutyl, butyl | Lake and river water | HPLC-UV/DF-µLPME | 1600–3500 | [38] |
| Methyl, ethyl, propyl, butyl | Wastewater, tap water | In-syringe MSPE/HPLC-PDA | 6–15 | This work |

Extraction methods: SPE: solid phase extraction, SBSE: stir-bar sorptive extraction, SPME: solid phase micro-extraction, DLE; dispersive liquid-liquid extraction, MSPD: matrix solid-phase dispersion, DF-µLPME: double-flow microfluidic based liquid phase micro-extraction, HF-MMLLE: hollow fiber-microporous membrane liquid-liquid extraction. Analytical methods: HPLC-CL: high performance liquid chromatography-chemiluminescence, UPLC: ultra-performance liquid chromatography-mass spectrometry, GC-MS: gas chromatography-mass spectrometry.

## 3.1. Analysis of Real Samples

The proposed method was also used in the preconcentration and determination of parabens in three kinds of real water samples (influent, effluent and tap water) that were analysed by HPLC-DAD after the in-syringe-SPME procedure. Under optimized conditions, the influent wastewater samples were spiked at two different levels (20 and 50 ng L$^{-1}$, $n = 3$) and analyzed according to the proposed procedure. At first the sample was analyzed without spiking. This was done in order to evaluate the levels on parabens in the original sample and the results of the analysis are shown in Table 6.

**Table 6.** Analysis of parabens in real samples (concentration in ng L$^{-1}$) using in-syringe MSPE/ HPLC-DAD, $n = 6$, degrees of freedom = 5.

| Samples | Methyl P | Ethyl P | Propyl | Butyl |
|---------|----------|---------|--------|-------|
| Influent 1 | 947 ± 3 | 168 ± 2 | 108 ± 1 | <LOQ |
| Influent 2 | 889 ± 2 | 267 ± 3 | 1988 ± 9 | 133 ± 3 |
| Effluent 1 | 392 ± 3 | 22.4 ± 0.9 | <LOQ | <LOQ |
| Effluent 2 | <LOQ | <LOQ | 1396 ± 5 | 132 ± 2 |
| Tap water | <LOQ | <LOQ | 629 ± 3 | <LOQ |

Figure 4 present the chromatogram of unspiked effluent water and a spiked sample after both were preconcentrated using in-syringe MSPE.

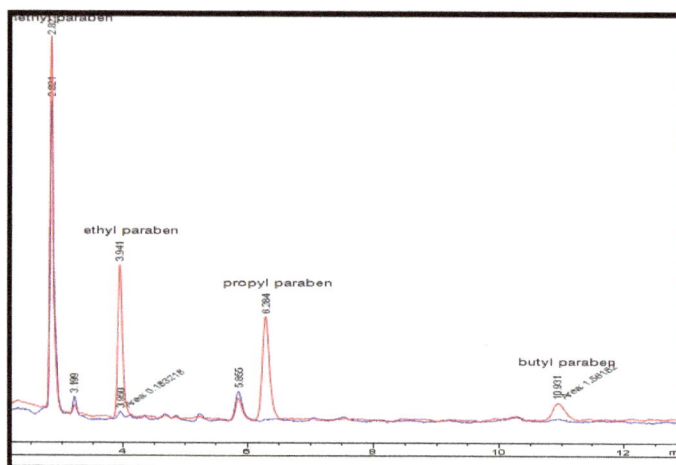

**Figure 4.** Chromatogram related to the extraction of target analytes (blue) wastewater (red) spiked wastewater at concentration level of 150 μg·L$^{-1}$ of analytes.

The percentage recoveries of paraben spiked at two levels were in the range of 97.5–99.4% with RSDs less than 5% (Table 7). The results demonstrated that the developed method can effectively preconcentrate and sensitively detect of trace amount of parabens in the presence of sample matrices.

**Table 7.** Validation of in-syringe MSPE for microextraction of parabens in spiked samples ($n = 4$).

| Added (ng L$^{-1}$) | Methyl | | Ethyl | | Propyl | | Butyl | |
|---|---|---|---|---|---|---|---|---|
| | Found (ng L$^{-1}$) | %Recovery | Found (ng L$^{-1}$) | %Recovery | Found (ng L$^{-1}$) | %Recovery | Found (ng L$^{-1}$) | %Recovery |
| 0 | 392 ± 3[a] | - | 22.4 ± 0.9 | | ND | | ND | - |
| 20 | 412 ± 5 | 98.0 ± 1.2 | 42.2 ± 1.2 | 99.0 | 19.5 ± 0.5 | 97.5 ± 2.6 | 19.6 ± 0.4 | 98.0 ± 2.0 |
| 50 | 442 ± 3 | 99.5 ± 0.8 | 71.6 | 98.4 | 49.6 ± 2.1 | 99.2 ± 4.2 | 49.7 ± 0.5 | 99.4 ± 4.0 |

[a] mean ± standard deviation ($n = 3$).

The in-syringe MSPE/HPLC-DAD method was successfully applied for preconcentration and determination of parabens in real samples and the analytical results are presented in Table 6. The results obtained by the current method were not significantly different from those obtained by reference method. These results obtained indicated the applicability of the in-syringe MSPE/ HPLC-DAD method for determination of parabens in environmental water samples

## 4. Material and Methods

### 4.1. Reagents and Standards

Methyl-, ethyl-, propyl- and butyl parabens were purchased from Sigma-Aldrich (St. Loius, MO, USA). HPLC grade solvents, including methanol and acetonitrile, were from Sigma-Aldrich (St. Louis, MO, USA). Stock solutions of parabens (10 mg $L^{-1}$) were prepared in ultra-pure water (Direct-Q® 3UV-R purifier system, Millipore, Merck, Germany) and stored in a freezer at 4 °C. The working standards were prepared by subsequent dilution of stocks.

### 4.2. Instrumentation

Chromatographic analysis was carried out by an Agilent HPLC 1200 infinity series system, equipped with photodiode array detector (Agilent Technologies, Waldbronn Germany). The chromatograms were recorded at 250 nm and 260 nm. An Agilent Zorbax Eclipse Plus C18 column (3.5 μm × 150 mm × 4.6 mm) (Agilent, Newport, CA, USA) was operated at an oven temperature of 25 °C. The mobile phase was a mixture of 30% water (Mobile phase A) and 70% methanol (mobile phase C). A flow rate of 1.00 mL min$^{-1}$ was used throughout the analysis. An OHAUS starter 2100 pH meter (Pine Brook, NJ, USA) was used for pH adjustments of the reagents and to measure the pH of samples.

The characterization of the adsorbents is crucial in their application for adsorption processes. The surface area ($S_{BET}$) and pore size distribution of the adsorbent was determined from $N_2$ adsorption-desorption isotherm using Brunauer, Emmett and Teller (BET) multipoint method using Surface Area and Porosity Analyzer (ASAP2020 V3. 00H, Micromeritics Instrument Corporation, Norcross, GA, USA). The pore volumes were calculated by the Barrerr-Joyner-Halenda method.

### 4.3. Methods

#### 4.3.1. Samples and Sample Collection

Both raw (influent) and treated (effluent) wastewater samples were used in this study. Sewage wastewater samples were collected in different points in the Daspoort wastewater treatment plant (Pretoria, Gauteng, South Africa). The samples were collected in pre-cleaned 500 mL glass bottles. After sampling the water samples were stored at 4 °C for a maximum of 1 week until being analysed.

#### 4.3.2. Preparation of CAC Adsorbent and Self-Made in-Syringe MSPE Device

Chitosan coated activated carbon adsorbent was synthesized according to the method described by Shariffard and colleagues [40]. Briefly, 5 g of activated carbon was treated with 0.2 mol $L^{-1}$ oxalic acid for 4 h. The activated carbon was then washed with deionised water after filtration and dried in an oven at 70 °C for 12 h. After this, chitosan was also prepared by adding 5 g of chitosan to 500 mL in 0.2 mol $L^{-1}$ oxalic acid solution under continuous stirring at 45–50 °C to form a viscous gel. Thereafter, the 5 g of the acid treated activated carbon was added slowly to the chitosan gel and stirred for 2 h at 45–50 °C. Then the activated carbon coated with chitosan beads were prepared by dropwise addition of the activated carbon gel into 0.7 M NaOH precipitation bath. The beads were filtered from NaOH bath and washed several times with deionised water to a neutral pH and dried in an air oven at 50 °C. The in-syringe SPE device [33] was prepared through slurry-packing CAC powders in the syringe-cartridge of with upper and lower filters. To discuss the methodology in detail, 81 mg CAC was added into 1 mL of deionized water and ultrasonicated to obtain a homogeneous dispersion.

Afterwards, the dispersion was transferred to the barrel, and the plunger was pushed slowly to remove the deionized water from dispersion. During this process, CAC particles were gradually deposited on the surface of lower filter. The upper filter was subsequently embedded and compacted tightly to obtain a stable packing bed.

### 4.3.3. In-Syringe SPE Procedure

For pre-concentration of parabens in model and real water samples, the self-made in-syringe-SPE device was first preconditioned with 1 mL methanol and 1 mL deionized water, respectively. After the conditioning step, the synthetic sample solution (100 µg $L^{-1}$) was passed through the device packed with adsorbent and then discharged from syringe cartridge though pushing the plunger. The estimated residence time of the sample solution in MSPE device was about 1 min. Subsequently, the plunger was drawn-pushed for several times in the air to remove the residual water from the device. The adsorbent was washed with 1 mL 5% methanol/water to remove the interferences. Finally, the parabens adsorbed on CAC were eluted with 800 µL methanol by aspirating and dispensing the eluent for two cycles.

The eluate was injected into HPLC-DAD system for qualitative and quantitative analysis. After each extraction, the adsorbent was successively washed with 1 mL methanol and 1 mL water for the next extraction. The optimization of extraction and pre-concentration conditions was performed using experimental design approach based central composite design (CCD). The factors investigated included mass of adsorbent, pH of sample solution and eluent volume (Table 8). The statistical analysis was achieved using Statistica software (Version 13, StatSoft, Inc., Tulsa, OK, USA).

**Table 8.** Factors and levels of experimental design.

| Variables | Low Level (−1) | Centre Point (0) | High Level (+) |
|---|---|---|---|
| Mass of Adsorbent (mg) | 25 | 72 | 100 |
| Sample pH | 4.0 | 6.6 | 9.0 |
| Eluent Volume (µL) | 200 | 720 | 1000 |

## 5. Conclusions

The study proposed a method which allowed for the simultaneous determination of trace levels of four common parabens in wastewater. Its procedure was based on an in-syringe-SPME pre-concentration technique followed by HPLC-DAD analysis of the parabens. This method was successfully applied to the analysis of wastewater collected from the Daspoort wastewater treatment plant (WWTP, Pretoria, Gauteng, South Africa). Compared to other methods reported in the literature, this technique offers simplicity, ease of use, cost effectiveness and lower consumption of organic solvents. It could effectively detect the most commonly used parabens (methyl and propyl paraben) throughout the study as the concentration of methyl paraben remained constant and propyl paraben was detected in all three analytes. Thus the adsorbent CAC was highly effective in the pre-concentration procedure for the analyses of the parabens

**Author Contributions:** G.P.M., A.M. and P.N.N. formulated the research idea; P.N.N. and G.P.M. designed the experiments; G.P.M. performed the actual experiments and data collection; G.P.M. and A.M. carried out the analysis of data and A.M. provided instrumental training; G.P.M. wrote the first draft of the manuscript and P.N.N. review and edited the final version of the manuscript; G.P.M., A.M. and P.N.N. collected real water samples.

**Funding:** This research was funded by National Research Foundation; grant No. 99720

**Acknowledgments:** This study was supported by the University of Johannesburg, South Africa (Department of Applied Chemistry, Centre for Nanomaterial Science Research) and National Research Foundation, South Africa.

**Conflicts of Interest:** The authors declare no conflict of interest.

## References

1. Juliano, C.; Magrini, G.A. Cosmetic ingredients as emerging pollutants of environmental and health concern. A Mini-Review. *Cosmetics* **2017**, *4*, 11. [CrossRef]
2. Zhong, Z.; Li, G. Current trends in sample preparation for cosmetic analysis. *J. Sep. Sci.* **2017**, *40*, 152–169. [CrossRef] [PubMed]
3. Brausch, J.M.; Rand, G.M. A review of personal care products in the aquatic environment: Environmental concentrations and toxicity. *Chemosphere* **2011**, *82*, 1518–1532. [CrossRef] [PubMed]
4. Blanco, E.; del Carmen Casais, M.; del Carmen Mejuto, M.; Cela, R. Combination of off-line solid-phase extraction and on-column sample stacking for sensitive determination of parabens and p-hydroxybenzoic acid in waters by non-aqueous capillary electrophoresis. *Anal. Chim. Acta* **2009**, *647*, 104–111. [CrossRef] [PubMed]
5. Hashemi, B.; Shamsipur, M.; Fattahi, N. Solid-phase extraction followed by dispersive liquid–liquid microextraction based on solidification of floating organic drop for the determination of parabens. *J. Chromatogr. Sci.* **2015**, *53*, 1414–1419. [CrossRef] [PubMed]
6. Ocaña-González, J.A.; Ramos-Payán, M.; Fernández-Torres, R.; Navarro, M.V.; Bello-López, M.Á. Application of chemiluminescence in the analysis of wastewaters—A review. *Talanta* **2014**, *122*, 214–222. [CrossRef] [PubMed]
7. Darbre, P.D.; Harvey, P.W. Paraben esters: Review of recent studies of endocrine toxicity, absorption, esterase and human exposure, and discussion of potential human health risks. *J. Appl. Toxicol.* **2008**, *28*, 561–578. [CrossRef] [PubMed]
8. Harvey, P.W.; Darbre, P. Endocrine disrupters and human health: Could oestrogenic chemicals in body care cosmetics adversely affect breast cancer incidence in women. *J. Appl. Toxicol.* **2004**, *24*, 167–176. [CrossRef] [PubMed]
9. Albero, B.; Sánchez-Brunete, C.; Miguel, E.; Pérez, R.A.; Tadeo, J.L. Determination of selected organic contaminants in soil by pressurized liquid extraction and gas chromatography tandem mass spectrometry with in situ derivatization. *J. Chromatogr. A* **2012**, *1248*, 9–17. [CrossRef] [PubMed]
10. Terasaki, M.; Makino, M. Determination of chlorinated by-products of parabens in swimming pool water. *Int. J. Environ. Anal. Chem.* **2008**, *88*, 911–922. [CrossRef]
11. Villar-Navarro, M.; del Carmen Moreno-Carballo, M.; Fernández-Torres, R.; Callejón-Mochón, M.; Bello-López, M.Á. Electromembrane extraction for the determination of parabens in water samples. *Anal. Bioanal. Chem.* **2016**, *408*, 1615–1621. [CrossRef] [PubMed]
12. Lee, M.R.; Lin, C.Y.; Li, Z.G.; Tsai, T.F. Simultaneous analysis of antioxidants and preservatives in cosmetics by supercritical fluid extraction combined with liquid chromatography–mass spectrometry. *J. Chromatogr. A* **2006**, *1120*, 244–251. [CrossRef] [PubMed]
13. Núñez, L.; Tadeo, J.L.; García-Valcárcel, A.I.; Turiel, E. Determination of parabens in environmental solid samples by ultrasonic-assisted extraction and liquid chromatography with triple quadrupole mass spectrometry. *J. Chromatogr. A* **2008**, *1214*, 178–182. [CrossRef] [PubMed]
14. Ferreira, A.M.C.; Möder, M.; Laespada, M.E.F. GC-MS determination of parabens, triclosan and methyl triclosan in water by in situ derivatisation and stir-bar sorptive extraction. *Anal. Bioanal. Chem.* **2011**, *399*, 945–953. [CrossRef] [PubMed]
15. Saraji, M.; Mirmahdieh, S. Single-drop microextraction followed by in-syringe derivatization and GC-MS detection for the determination of parabens in water and cosmetic products. *J. Sep. Sci.* **2009**, *32*, 988–995. [CrossRef] [PubMed]
16. Levchyk, V.M.; Zui, M.F. Gas Chromatographic determination of parabens after derivatization and dispersive microextraction. *Fr.-Ukr. J. Chem.* **2015**, *3*, 72–79. [CrossRef]
17. Chen, C.W.; Hsu, W.C.; Lu, Y.C.; Weng, J.R.; Feng, C.H. Determination of parabens using two microextraction methods coupled with capillary liquid chromatography-UV detection. *Food Chem.* **2018**, *241*, 411–418. [CrossRef] [PubMed]
18. Xiu-Qin, L.; Chao, J.; Wei, Y.; Yun, L.; Min-Li, Y.; Xiao-Gang, C. UPLC-PDAD analysis for simultaneous determination of ten synthetic preservatives in foodstuff. *Chromatographia* **2008**, *68*, 57–63. [CrossRef]

19. Geara-Matta, D.; Lorgeoux, C.; Rocher, V.; Chebbo, G.; Moilleron, R. Contamination of wastewater by endocrine disruptors in France: Analytical development for triclosan, triclocarban and parabens. *Tech. Sci. Method* **2011**, *10*, 17–24.

20. Márquez-Sillero, I.; Aguilera-Herrador, E.; Cárdenas, S.; Valcárcel, M. Determination of parabens in cosmetic products using multi-walled carbon nanotubes as solid phase extraction sorbent and corona-charged aerosol detection system. *J. Chromatogr. A* **2010**, *1217*, 1–6. [CrossRef] [PubMed]

21. Piao, C.; Chen, L.; Wang, Y. A review of the extraction and chromatographic determination methods for the analysis of parabens. *J. Chromatogr. B* **2014**, *969*, 139–148. [CrossRef] [PubMed]

22. Farajzadeh, M.A.; Djozan, D.; Bakhtiyari, R. Use of a capillary tube for collecting an extraction solvent lighter than water after dispersive liquid–liquid microextraction and its application in the determination of parabens in different samples by gas chromatography—Flame ionization detection. *Talanta* **2010**, *81*, 1360–1367. [CrossRef] [PubMed]

23. Huang, Y.; Peng, J.; Huang, X. One-pot preparation of magnetic carbon adsorbent derived from pomelo peel for magnetic solid-phase extraction of pollutants in environmental waters. *J. Chromatogr. A* **2018**, *1546*, 28–35. [CrossRef] [PubMed]

24. Jain, R.; Mudiam, M.K.R.; Chauhan, A.; Ch, R.; Murthy, R.C.; Khan, H.A. Simultaneous derivatisation and preconcentration of parabens in food and other matrices by isobutyl chloroformate and dispersive liquid–liquid microextraction followed by gas chromatographic analysis. *Food Chem.* **2013**, *141*, 436–443. [CrossRef] [PubMed]

25. Ramírez, N.; Borrull, F.; Marcé, R.M. Simultaneous determination of parabens and synthetic musks in water by stir-bar sorptive extraction and thermal desorption-gas chromatography-mass spectrometry. *J. Sep. Sci.* **2012**, *35*, 580–588. [CrossRef] [PubMed]

26. González-Mariño, I.; Quintana, J.B.; Rodríguez, I.; Schrader, S.; Moeder, M. Fully automated determination of parabens, triclosan and methyl triclosan in wastewater by microextraction by packed sorbents and gas chromatography–mass spectrometry. *Anal. Chim. Acta* **2011**, *684*, 59–66. [CrossRef] [PubMed]

27. Cabaleiro, N.; De La Calle, I.; Bendicho, C.; Lavilla, I. An overview of sample preparation for the determination of parabens in cosmetics. *TrAC Trends Anal. Chem.* **2014**, *57*, 34–46. [CrossRef]

28. Wen, Y.; Chen, L.; Li, J.; Liu, D.; Chen, L. Recent advances in solid-phase sorbents for sample preparation prior to chromatographic analysis. *TRAC Trends Anal Chem.* **2014**, *59*, 26–41. [CrossRef]

29. Farajzadeh, M.A.; Yadeghari, A.; Khoshmaram, L. Combination of dispersive solid phase extraction and dispersive liquid-liquid microextraction for extraction of some aryloxy pesticides prior to their determination by gas chromatography. *Microchem. J.* **2017**, *131*, 182–191. [CrossRef]

30. Sajid, M.; Basheer, C. Layered double hydroxides: Emerging sorbent materials for analytical extractions. *TrAC Trends Anal. Chem.* **2016**, *75*, 174–182. [CrossRef]

31. Rodríguez-González, N.; González-Castro, M.J.; Beceiro-González, E.; Muniategui-Lorenzo, S.; Prada-Rodríguez, D. Determination of triazine herbicides in seaweeds: Development of a sample preparation method based on matrix solid phase dispersion and solid phase extraction clean-up. *Talanta* **2014**, *121*, 194–198. [CrossRef] [PubMed]

32. Vázquez, P.P.; Mughari, A.R.; Galera, M.M. Application of solid-phase microextraction for determination of pyrethroids in groundwater using liquid chromatography with post-column photochemically induced fluorimetry derivatization and fluorescence detection. *J. Chromatogr. A* **2008**, *1188*, 61–68. [CrossRef] [PubMed]

33. Zhang, X.; Wang, P.; Han, Q.; Li, H.; Wang, T.; Ding, M. Metal-organic framework based in-syringe solid-phase extraction for the on-site sampling of polycyclic aromatic hydrocarbons from environmental water samples. *J. Sep. Sci.* **2018**. [CrossRef] [PubMed]

34. Dimpe, K.M.; Mpupa, A.; Nomngongo, P.N. Microwave assisted solid phase extraction for separation preconcentration sulfamethoxazole in wastewater using tyre based activated carbon as solid phase material prior to spectrophotometric determination. *Spectrochim. Acta Part A Mol. Biomol. Spectrosc.* **2018**, *188*, 341–348. [CrossRef] [PubMed]

35. Sharififard, H.; Nabavinia, M.; Soleimani, M. Evaluation of adsorption efficiency of activated carbon/chitosan composite for removal of Cr (VI) and Cd (II) from single and bi-solute dilute solution. *Adv. Environ. Technol.* **2017**, *2*, 215–227. [CrossRef]

36. Nomngongo, P.N.; Ngila, J.C.; Msagati, T.A.; Moodley, B. Chemometric optimization of hollow fiber-liquid phase microextraction for preconcentration of trace elements in diesel and gasoline prior to their ICP-OES determination. *Microchem. J.* **2014**, *114*, 141–147. [CrossRef]

37. Khodadoust, S.; Sadeghi, H.; Pebdani, A.A.; Mohammadi, J.; Salehi, A. Optimization of ultrasound-assisted extraction of colchicine compound from Colchicum haussknechtii by using response surface methodology. *J. Saudi Soc. Agric. Sci.* **2017**, *16*, 163–170. [CrossRef]

38. Ramos-Payan, M.; Maspoch, S.; Llobera, A. A simple and fast Double-Flow microfluidic device based liquid-phase microextraction (DF-µLPME) for the determination of parabens in water samples. *Talanta* **2017**, *165*, 496–501. [CrossRef] [PubMed]

39. Luiz Oenning, A.; Lopes, D.; Neves Dias, A.; Merib, J.; Carasek, E. Evaluation of two membrane-based microextraction techniques for the determination of endocrine disruptors in aqueous samples by HPLC with diode array detection. *J. Sep. Sci.* **2017**, *40*, 4431–4438. [CrossRef] [PubMed]

40. Regueiro, J.; Becerril, E.; Garcia-Jares, C.; Llompart, M. Trace analysis of parabens, triclosan and related chlorophenols in water by headspace solid-phase microextraction with in situ derivatization and gas chromatography–tandem mass spectrometry. *J. Chromatogr. A* **2009**, *1216*, 4693–4702. [CrossRef] [PubMed]

41. Ebrahimpour, B.; Yamini, Y.; Esrafili, A. Emulsification liquid phase microextraction followed by on-line phase separation coupled to high performance liquid chromatography. *Anal. Chim. Acta* **2012**, *751*, 79–85. [CrossRef] [PubMed]

42. González-Mariño, I.; Quintana, J.B.; Rodríguez, I.; Cela, R. Simultaneous determination of parabens, triclosan and triclocarban in water by liquid chromatography/electrospray ionisation tandem mass spectrometry. *Rapid Commun. Mass Spectrom.* **2009**, *23*, 1756–1766. [CrossRef] [PubMed]

43. Feizi, N.; Yamini, Y.; Moradi, M.; Karimi, M.; Salamat, Q.; Amanzadeh, H. A new generation of nano-structured supramolecular solvents based on propanol/gemini surfactant for liquid phase microextraction. *Anal. Chim. Acta* **2017**, *953*, 1–9. [CrossRef] [PubMed]

44. Molins-Delgado, D.; Díaz-Cruz, M.S.; Barceló, D. Ecological risk assessment associated to the removal of endocrine-disrupting parabens and benzophenone-4 in wastewater treatment. *J. Hazard. Mater.* **2016**, *310*, 143–151. [CrossRef] [PubMed]

*molecules*

MDPI

*Review*

# Solid-Phase Extraction of Polar Benzotriazoles as Environmental Pollutants: A Review

**Ida Kraševec and Helena Prosen ***

Faculty of Chemistry and Chemical Technology, University of Ljubljana, Večna pot 113, SI-1000 Ljubljana, Slovenia; ida.krasevec@fkkt.uni-lj.si
* Correspondence: helena.prosen@fkkt.uni-lj.si; Tel.: +386-1-479-8556

Received: 6 September 2018; Accepted: 26 September 2018; Published: 29 September 2018

**Abstract:** Polar benzotriazoles are corrosion inhibitors with widespread use; they are environmentally characterized as emerging pollutants in the water system, where they are present in low concentrations. Various extraction methods have been used for their separation from various matrices, ranging from classical liquid–liquid extractions to various microextraction techniques, but the most frequently applied extraction technique remains the solid-phase extraction (SPE), which is the focus of this review. We present an overview of the methods, developed in the last decade, applied for the determination of benzotriazoles in aqueous and solid environmental samples. Several other matrices, such as human urine and plant material, are also considered in the text. The methods are reviewed according to the determined compounds, sample matrices, cartridges and eluents used, extraction recoveries and the achieved limits of quantification. A critical evaluation of the advantages and drawbacks of the published methods is given.

**Keywords:** benzotriazoles; solid-phase extraction; environmental samples

## 1. Introduction

Solid-phase extraction (SPE) is one of the most frequently used techniques for sample preparation in environmental analysis, not only for well-known organic pollutants, but also for contaminants of emerging concern. The latter group includes polar benzotriazoles, which are the focus of this review. We aim to present an overview of the SPE methods applied for the determination of benzotriazoles in aqueous and solid environmental samples, as well as in several other matrices. The literature reviewed in this work has been published in the last decade (2008–2018).

In general, SPE is a procedure in which the dissolved analytes are put into contact with a solid phase, where they are retained due to the interactions with the sorbent. Ideally, the matrix of the sample solution remains unaffected by the sorbent, and solely the analytes are extracted from the sample. The selectivity of the extraction is mainly determined by the sorbent, which can be chosen among reversed-phase ($C_{18}$), normal-phase (silica), ion exchange and mixed-mode (polymeric) phases. Moreover, in the many decades of SPE use, various forms of sorbents have been developed, for example sorbents in the shape of disks, cartridges and well plates. The principal difference between these configurations is their capacity, which is directly related to the mass of the contained sorbents. In environmental research, cartridges are the prevalent form of SPE sorbents.

Generally, an SPE method consists of four steps, usually performed with the aid of a vacuum manifold. The cartridge is first conditioned with an organic solvent and pH-adjusted water solution. This minimizes the surface tension of the sorbent (especially important in reversed-phase sorbents) and enables interactions with the desired form of sorbents (protonated or de-protonated in ion exchange and mixed-mode sorbents) [1]. Secondly, the sample is loaded onto the sorbent and a washing or rinsing step is performed. This facilitates the removal of weakly bound interferences (matrix components) and strengthens the analyte–sorbent interactions. Finally, the sorbent is dried under vacuum or under nitrogen flow and the analytes are eluted with an organic solvent optimized to sever the analyte–sorbent interactions (usually pH-adjusted in ion exchange and mixed-mode sorbents).

## 2. Benzotriazoles

Benzotriazoles are heterocyclic compounds, with the basic structure of a benzene ring fused to a triazole ring. Non-polar benzotriazoles, usually hydroxyphenyl derivatives, are used as UV stabilizers in plastic materials and personal care products [2]. Polar benzotriazoles, including the basic benzotriazole and the methyl-, hydroxy- and chloro- derivatives, are used as corrosion inhibitors in antifreeze liquids, various industrial systems (cooling, braking, cutting and metal-working fluids), aircraft de-icing fluids and household dishwashing detergents [2–4]. The most commonly used are the benzotriazole (BTZ) and a mixture of 4- and 5-methyl-benzotriazole (MBZ) isomers, also known as tolyltriazole.

In this review, we focus on the polar compounds listed in Table 1. Due to a high aqueous mobility stemming from their polarity (observable as $\log K_{ow}$ in Table 1) they are ubiquitously present in the hydrosphere, having been found in snow, groundwater, rivers, lakes, seas, treated and untreated wastewater, recycled water and also in drinking water [2,5–11]. Furthermore, they have been also determined in air, soil, sediments, house dust and textiles [12–16], in human urine, amniotic fluid and plants [11,17,18].

**Table 1.** Table of the reviewed compounds, their abbreviations and selected chemical properties ($\log K_{ow}$, $pK_a$: predicted values from the Scifinder database).

| Abbreviation | Full Name | $\log K_{ow}$ | $pK_a$ |
|---|---|---|---|
| BTZ | 1*H*-benzotriazole | 1.44 | 8.38 |
| 1MBZ | 1-methyl-1*H*-benzotriazole | 1.08 | – |
| 2MBZ | 2-methyl-2*H*-benzotriazole | 1.59 | – |
| 4MBZ | 4-methyl-1*H*-benzotriazole | 1.82 | 8.74 |
| 5MBZ | 5-methyl-1*H*-benzotriazole | 1.98 | 8.74 |
| DMBZ | 5,6-dimethyl-1*H*-benzotriazole | 2.28 | 8.92 |
| 24DMBZ | 2,4-dimethyl-2*H*-benzotriazole | 1.96 | – |
| ClBZ | 5-chloro-1*H*-benzotriazole | 2.13 | 7.46 |
| NBZ | 5-amine-1*H*-benzotriazole | 0.40 | 9.61 |
| 1OHBZ | 1-hydroxy-benzotriazole | 0.69 | 7.39 |
| 4OHBZ | 4-hydroxy-benzotriazole | 0.80 | 7.25 |

The main sources of these pollutants appear to be airport run-off and industrial and municipal wastewater [2,5,8], as benzotriazoles are not readily biodegradable. According to different sources, only 13–62% of BTZ and 11–72% of MBZ are removed in public wastewater treatment plants [2,5,19] due to their long half-lives [20], with 5MBZ being much more susceptible to degradation than 4MBZ. The compounds can also pass through common steps in the preparation of drinking water [21], but can be removed from recycled water with additional treatment steps, such as reverse osmosis after ultrafiltration [22], ozonization [23] or photodegradation with UV/$H_2O_2$ [24]. Degradation pathways in activated sludge are not well known, but the suggested metabolites of aerobic digestion appear to include phenol, phthalic acid, 1-methyl benzotriazole, 4-methoxy-1*H*-benzotriazole and 5-methoxy-1*H*-benzotriazole for BTZ, 2,5-dimethyl benzoxazole and benzotriazole for 5MBZ and 5-chloro-2-methyl benzoxazole and benzotriazole for ClBZ [20].

The benzotriazoles have been shown to be toxic to aquatic bacteria, plants and invertebrates [25]. BTZ is also a possible carcinogen [6] and expresses anti-estrogenic activity [26].

As these compounds are classified as emerging pollutants, the limits for their acceptable concentrations in the environment are not generally set. The only exceptions we found in the literature are the limits of 5 mg/m$^3$ BTZ in workplace air and 0.1 mg/L BTZ in drinking water from Russia, cited by Pervova et al. [27], and the limit of 7 ng/L 5MBZ in drinking water of Australia, reported by Janna et al. [6].

Due to low environmental concentrations, the determination methods use sensitive chromatographic techniques, such as liquid or gas chromatography with mass spectrometric detection (LC-MS, GC-MS) in concert with various extraction techniques. Along with the classical SPE, newer techniques have also been applied for their determination, for example QuEChERS (Quick, Easy, Cheap, Effective, Rugged and Safe extraction) [28], stir-bar sorptive extraction (SBSE) [29], solid-phase microextraction (SPME) [30,31] and dispersive liquid–liquid microextraction (DLLME) with optional concurrent derivatization [32,33]. In some cases, determination without pre-concentration has also been reported for samples with higher analyte concentrations [2,22], but the vast majority of the studies make use of SPE to achieve a pre-concentration of analytes and a reduction of matrix effects.

For readers with further interest in benzotriazoles, we suggest a dedicated review on toxicity and occurrence of BTZ and 5MBZ, which has been published by Alotaibi et al. [34], and a review of available analytical methods, published by Herrero et al. [35]. In contrast to these works, this review focuses only on SPE as the sample preparation technique for determination of polar benzotriazoles and also considers the research, published in the years following the publication of the previous reviews.

## 3. Analytical Techniques for the Determination of Polar Benzotriazoles

In environmental chemistry, the determination of benzotriazoles is commonly done with chromatographic methods. Due to the compounds' polarity, low volatility and occurrence in low concentrations, the most commonly chosen techniques are LC for separation and MS for sensitive detection and identification. Occasionally, UV/VIS or fluorescence detection [32,36] have also been used, but noticeably higher limits of quantification (LOQ) have been achieved in those cases.

In the majority of cases, separation has been performed on various $C_{18}$ columns [4,10,11,21,37–39], although columns with phenyl stationary phase have also been used [2,8,22]. Usually, the mobile phase has initially a high water content and is then gradually adjusted to raise the ratio of organic solvent, this solvent being either methanol (MeOH) [10,40], acetonitrile (ACN) [22,32] or both [38,41]. The critical problem of the LC is the separation of 4- and 5-methyl-benzotriazole isomers, due to their similar retention. It has been claimed that the use of ACN is necessary for their separation [35] and in the cases where both solvents were tested, better chromatographic profiles and lower retention times [32] or better peak heights and shapes [39] were achieved when using ACN. When LC is coupled with MS, formic or acetic acid is added to the mobile phase to increase the ionization efficiency of the analytes [4,8,21,41].

Ionization is almost invariably performed in the positive mode of electrospray ionization (ESI+), since atmospheric pressure chemical ionization (APCI) was proven to be much less sensitive, resulting in higher limits of detection (LOD) [22]. To achieve high sensitivity, tandem MS (MS/MS) or high resolution MS (HRMS) are commonly used. MS/MS is usually performed with a triple quadrupole analyser in selected reaction monitoring (SRM) mode, with two transitions being recorded for each analyte to satisfy the requirements for quantification and identification [10,42]. For HRMS, performed with Orbitrap [21,37,38] or time-of-flight (TOF) [22] analysers, only a precursor and a product ion are required for a proper identification of the analytes. The main drawback of LC-MS methods is the appearance of the matrix effect, which is the suppression or enhancement of analyte ionization due to the sample matrix. This effect depends on the matrix composition of the injected sample and the concentration level of analytes, and is often more pronounced at lower concentration levels. It is one of the main reasons for using intensive sample clean-up procedures. Other methods of circumventing this problem include the use of matrix-matched calibration or an isotopically labelled internal standard [35].

For quantification of benzotriazoles in environmental samples, external calibration [4,11,21,38,43], matrix-matched calibration [17,39,42] and the standard addition method [8,23] have been used to date. External calibration is generally used in cases of simple samples and very low matrix effects, while standard addition and matrix-matched calibration are used in the case of more complex samples, since they compensate for some of the matrix effects.

In comparison to LC, GC is used less often. Although polar benzotriazoles are quite non-volatile, they can be determined either with derivatization (trimethylsulfonium hydroxide (TMSH), acetylation [33,44,45]) or without [7,46], to reduce the complexity of the method [47]. The stationary phase used is 5%-phenyl-methylpolysiloxane or equivalent [5,7,45,46,48–50], except in rare cases, for example [51], where the authors tested various ionic liquids as stationary phases and reported faster separation and better peak shape than with conventional columns. A two-dimensional GC × GC-TOF method has been developed by Jover et al. [47], who separated co-eluted analytes and achieved LODs similar to LC MS/MS.

Almost invariably, MS [5,45,49,50,52] or MS/MS [7,46] detectors are used after separation. In one case, isotope ratio MS was also performed, but the authors reported difficulties with low sensitivity, low volatility, and interactions of analytes with the exposed metal components of the instrument, demanding instrument modification or derivatization of the analytes [44]. The main drawbacks of GC-MS appear to be poor retention on non-polar columns [51], peak tailing or peak disappearance on polar columns and low sensitivity due to low *m/z* of the target ions [47].

## 4. Solid-Phase Extraction of Benzotriazoles from Aqueous Samples

As mentioned before, SPE is the prevalent extraction technique for polar benzotriazoles in environmental samples due to its high pre-concentration factors (large sample volume), smaller amounts of solvent used (in comparison to liquid–liquid extraction, LLE), practical simplicity and the possibility of automatization. The methods developed in the last decade for determination of benzotriazole pollutants in aqueous environmental samples have been collected in Table 2. In addition to the sorbent phase used, several other parameters are presented. The choice of the elution solvent can depend on recoveries and the selected analytical technique (for example volatile solvents for GC), although the elution solvent is often evaporated and the sample reconstituted in a suitable solvent (this can also serve as an additional concentrating step). Where possible, recoveries and LOQs were added as quality control parameters, as reported by the authors. Usually, the recoveries were calculated comparing the instrument response of a post-extraction spiked sample to a pre-extraction spiked sample, either in river waters or in effluent wastewater (when the former was not available). LOQs are given as reported by the authors, but it is necessary to consider that they also depend on the analytical techniques used for determination. Some of the studies reported in the Table 2, are discussed here more in-depth, as their authors presented novel research and optimization of SPE procedures (pH of the sample solution, reduction of matrix effect, sorbent phases used, etc.).

It is worth noting that in most of the reviewed analytical methods, benzotriazoles are not the only compounds being determined. Often they are the subject of a combined analysis with benzothiazoles, benzenesulfonamides or UV stabilizers [4,7,38,47], or some of them are targets of screening methods that include tens or hundreds of other micropollutants (pesticides, pharmaceuticals, personal care products, etc.) [53,54]. Naturally, this leads to the development of methods that are not perfectly optimized for all the determined compounds, as can be observed for example in recoveries of polar benzotriazoles ranging from 22 to 112% in [54]. In some of the reviewed papers, there is also very little to no emphasis on method optimization, since they focus more on the application and interpretation of concentrations and their changes in various environmental samples [5,6,52,55–58].

Considering the relatively high polarity and weakly acidic character of the polar benzotriazoles (Table 1), the relevant extraction phases for their determination include polymeric balanced polar/non-polar (HLB, Strata-X, Bond Elut PPL), strong anion (Oasis MAX, Strata-X-A) and cation exchange (Oasis MCX, Strata-X-C) sorbents (Oasis series by Waters, Strata series by Phenomenex, Bond Elut by Agilent), while $C_{18}$ and charcoal based sorbents have been shown to be much less effective in retention than HLB [7]. As can be seen in Table 2, the most widely used cartridge sorbent is the HLB polymeric sorbent, a N-vinylpyrrolidone divinylbenzene copolymer with good retention for non-polar and neutral polar compounds. In most cases [5–7,10,21,44,47,52] the samples were acidified to pH 2–3. Some authors claim that the sample pH had great effect on recovery [10,21], shifting from less than 50% at pH 7, to 79–110% at pH 2 [7]. This is countered by authors claiming that pH had no significant effect on recovery, but the ionization suppression was higher at pH 3 than at pH 7 [4], or that the effect of pH on BTZ recovery was very low for HLB and various other sorbents (Oasis MAX, MCX, WAX, WCX, Strata-X, X-A, X-AW, X-C, X-CW) [23].

In the cases where the research focuses on the extraction method, the main aim of development is usually a reduction of matrix effects, not higher recoveries. One of the steps that has been taken to achieve additional clean-up of the samples, was the use of a Florisil filled cartridge directly connected to the HLB cartridge. In this way the SPE eluent came directly into contact with the Florisil sorbent (magnesium silicate) which adsorbed the interferences, thereby reducing the matrix effect in spiked wastewater from $-65$ to $-40\%$ without Florisil to $-5$ to $-15\%$ with Florisil [4].

A mixed-mode sorbent, which could also reduce matrix effects, has been applied for benzotriazoles for the first time by Carpinteiro et al. [8]. Oasis MAX cartridges, containing strong anion exchange and reversed-phase sorbent, were eluted with the same organic solvent mixture (MeOH/ACO) as Oasis HLB. In comparison to HLB, MAX cartridges produced cleaner extracts and lower matrix effects, which lead to the conclusion that interfering matrix components (for example humic, fulvic acids) remained retained with anionic interaction. Although the sample pH was not modified, the recoveries ranged 96–102% in river water and 82–86% in effluent wastewater, but only 50–55% in influent wastewater. These recoveries correlated with the reported matrix effects: none for river water, small suppression for effluent wastewater, and substantial suppression for influent wastewater, which was also the most complex of the investigated matrices.

The mechanisms of sorption in the mixed-mode phases have been studied by Salas et al. [38] in both Oasis MAX and MCX cartridges. The authors optimized extraction procedures for both sorbents, each including two rinsing steps. The first rinsing step consisted of pH-adjusted water, to wash off weakly adsorbed hydrophilic compounds and increase ionic interactions, while in the second step they used MeOH to rinse neutral compounds retained by hydrophobic interactions. With MCX cartridges, the analytes remained adsorbed during washing, even though the sample pH was adjusted to pH 3 and the analytes were in their neutral forms. When increasing the volume of MeOH used for rinsing, out of the 5 benzotriazoles investigated only ClBZ washed off, indicating the weakest adsorption. In MAX they noticed no differences in retention at pH 7 and 11.5, although the analytes were de-protonated at the higher pH. This led to conclusion that the mechanism of retention in both phases is through induced dipole-ionic interactions, possible due to the delocalized electron density and electronegative elements in the benzotriazole rings. Rinsing lowered the amount of interferences in the extracts and resulted in very low ionization suppression in influent wastewater extracts, while achieving high recoveries with both cartridges (61–92%).

Although higher recoveries have been reported to be achieved with HLB (more than 90%) than with MAX at pH 10 (20–64%) in wastewater [4], these examples state that to the contrary, high recoveries can be achieved also with mixed-mode cartridges.

An innovative tandem SPE method for determination of analytes of various properties (BTZ among them) has been developed by Deeb and Schmidt [23]. The Oasis MAX and MCX cartridges were connected consecutively for the sample loading and then separated for washing and elution. All the analytes in the combined extract had recoveries more than 90%, while the matrix effects were lower than 13% for all matrices, including wastewater, implying that good clean-up was achieved for all samples.

Layered 'mixed bed' cartridges have also been used for determination of a broad range of compounds, including benzotriazoles [54,56,59–61], with the aim of extracting hydrophobic, hydrophilic, anionic and cationic compounds in one step. The lab-prepared layered cartridges contained a layer of Oasis HLB sorbent on the top, to adsorb compounds with non-specific interactions, and a layer of Strata-X-AW, Strata-X-CW and Isoelute ENV+ mixture on the bottom, to adsorb the remaining compounds either by ionic or hydrophilic interactions. The eluents used were acidified and alkaline organic solvents, MeOH, ACN or ethyl acetate (EA). Osorio et al. [54] reported achieving lower matrix effects with the mixed bed cartridges in comparison to the HLB sorbent (which is critical for reducing the amount of co-elutions in the LC) and higher recoveries in comparison to the MAX sorbent.

**Table 2.** SPE methods reported for polar benzotriazoles in aqueous environmental samples (eff–effluent, ww–wastewater, ND–no data) .

| Compounds | Sample Matrices | Cartridge Type, mg | Elution Solvent | Recovery | LOQ (ng/L) | Reference |
|---|---|---|---|---|---|---|
| BTZ, 4MBZ, 5MBZ, DMBZ, 1OHBZ, ClBZ | tap, surface, effluent | Oasis HLB 500 | 3 × 2.5 mL ACN/MeOH (1:1) | 57–125% | 2–11 | [21,37] |
| BTZ, 4MBZ, 5MBZ, DMBZ | river, influent, effluent | Strata-X | 5 mL EA | 85–115% | 31–99 | [47] |
| BTZ, 4MBZ, 5MBZ | river | Bond Elut PPL 200 | 2 mL ACN/MeOH (1:1) | 49–79% | 6–15 | [5] |
| BTZ, 1MBZ | groundwater, river | Oasis HLB 200 | 6 ml MeOH | 47–56% | 1 | [62,63] |
| BTZ, 4MBZ, 5MBZ | groundwater | Oasis HLB 200 | 7 mL MeOH/ACO (6:4) | 95–113% | 50 | [2] |
| BTZ, 5MBZ | tap, river, influent, effluent | Oasis HLB 500 | 5 mL 3% MeOH in DCM | 101–118% (eff) | 0.2 | [6] |
| BTZ, 5MBZ | tap, groundwater, influent, effluent | Oasis HLB 500 | 3 × 2 mL DCM/MeOH (1:1) | 79–108% | 9.8–36.7 | [7,46,48] |
| BTZ, 5MBZ | river, sea | Oasis HLB 500 | 15 mL MeOH | 69 ± 10% (sea) | 0.4–1.2 | [40] |
| BTZ, 5MBZ, DMBZ | river, influent, effluent | Oasis MAX 150 | 5 mL MeOH / ACN (7:3) | 87–99% | 2–5 | [8] |
| BTZ, 5MBZ, DMBZ | influent, effluent | Strata-X 100 | 10 mL hexane/EA (1:1) | 78–98% (ww) | 60–810 | [51] |
| BTZ, 1MBZ, 4MBZ, 5MBZ | groundwater, river, effluent | Oasis HLB (10) + Strata-X-AW(1.4)/ Strata-X-CW(1.4)/ Isoelute ENV+(2.1) | 0.1% HCOOH in MeOH | 94–124% | 0.2–4.2 | [59] |
| BTZ, 5MBZ, DMBZ, 1OHBZ | river, effluent | Strata-X 200 | 2 × 5 mL MeOH/ACN (1:1) | 73–104% (ww) | 0.42–10 | [39] |
| BTZ, 4MBZ, 5MBZ, DMBZ, ClBZ | river, influent, effluent | Oasis HLB 150 + Florisil 500 | 2 × 3 mL MeOH | 85–93% | 2 | [4] |
| BTZ, 5MBZ | sea | Oasis HLB 200 | 6 mL MeOH | 67–75% | 0.11–0.17 | [53] |
| BTZ, 5MBZ | effluent | Oasis HLB 200 | 6 mL MeOH | 47–56% | 40 (eff) | [64] |
| BTZ, 1MBZ, 4MBZ, 5MBZ | tap, effluent | Oasis HLB 200 | 10 mL EA | 80–93% | ND | [44] |
| BTZ, 4MBZ, 5MBZ | airport runoff | Strata C-18 E 500 | 40 mL DCM | 68–102% | 0.3–10 | [49] |
| BTZ, 1MBZ, 4MBZ, 5MBZ, 1OHBZ, 4OHBZ | recycled water | Oasis HLB 200 + Strata-X-AW(100)/ Strata-X-CW(100)/ Isoelute ENV+(150) | 8 mL EA/MeOH/NH₃ + 4 mL EA/MeOH/ HCOOH | ND | 1-50 | [60] |
| BTZ, 4MBZ, 5MBZ, DMBZ, ClBZ | wastewater | Strata-X 200 | 10 mL MeOH/ACN (1:1) | 36–85% | 52–376 | [65,66] |
| BTZ, 5MBZ | river, influent, effluent | PLRP-s | ACN/H₂O/HCOOH (24.9:74.9:0.2) | 85–100% (eff) | 0.8–1.1 | [42,67] |
| BTZ | effluent | Bond Elut PPL 200 | ND | ND | 50 | [55] |
| BTZ | tap, river, effluent | Oasis MAX (100) + Oasis MCX (100) | separate elution: MAX 6 mL MeOH/EA /HCOOH (69:29:2), MCX 6 mL MeOH/EA/NH₃ (67.5:27.5:5) | 93 ± 4% | 5.7 | [23] |
| 2MBZ, 24DMBZ | groundwater, river | Merck EN 200 | 10 mL DCM | ND | 3.3–6.7 | [50] |
| BTZ, 4MBZ, 5MBZ, DMBZ, ClBZ | river, influent, effluent | Oasis MCX 500, Oasis MAX 500 | MCX: 5 mL 5% NH₃ in MeOH, MAX: 5 mL 5% HCOOH in MeOH | MCX: 87–105%, MAX: 75–92% | MCX: 5–17, MAX: 5–21 | [38] |

**Table 2.** *Cont.*

| Compounds | Sample Matrices | Cartridge Type, mg | Elution Solvent | Recovery | LOQ (ng/L) | Reference |
|---|---|---|---|---|---|---|
| BTZ, 5MBZ | groundwater | PLRP-s | ACN/H$_2$O/HCOOH (24.9:74.9:0.2) | | | [68] |
| BTZ, 4MBZ, 5MBZ, 1OHBZ, CIBZ | tap, lake, effluent | Poly-Sery HLB 60 | 2 mL MeOH | 89–103% | 0.4–4.8 | [10] |
| BTZ | tap, recycled water | Oasis HLB 500 | 5 mL MeOH + 5 mL MTBE/MeOH (9:1) | | | [11] |
| BTZ, MBZ, DMBZ, CIBZ | influent, effluent | Oasis MCX 60 | 6 mL MeOH + 4 mL ACN | 72–102% | 500–2000 | [58] |
| BTZ, MBZ | river | Oasis HLB 200 | 10 mL EA + 10 mL 0.1% NH$_3$ in MeOH | 31–72% | 0.53–0.66 | [41] |
| BTZ, DMBZ, CIBZ, NBZ | tap, swimming pool | Poly-Sery HLB 200 | 6 mL MeOH | 95–105% | 1.5–15 | [69] |
| BTZ, MBZ | influent, effluent | Oasis HLB 200 | 5 mL 5% HCOOH in MeOH + 5 mL EA | 108–154% | 10–20 | [70] |
| BTZ, 5MBZ | river, influent, effluent, aquaculture ponds | Oasis HLB | 3 mL MeOH + 3 mL diethyl ether/MeOH (1:1) | 91–104% | 10 | [71,72] |
| BTZ | river | Supel-Select HLB 200 | 6 mL MeOH/ACO/EA (2:2:1) | 94% | ND | [73] |
| BTZ, MBZ | effluent | Oasis HLB 9 + Strata-X-AW(2.6)/ Strata-X-CW(2.6)/ Isoelute ENV+(3.8) | 0.1% HCOOH in MeOH | 97–102% | 2–5 | [61] |
| BTZ, 5MBZ, CIBZ, DMBZ | river | Poly-Sery HLB 200 | 6 mL MeOH | 80–110% | 0.4–1.5 | [74] |
| BTZ, 4MBZ, 5MBZ, CIBZ, DMBZ | river | Oasis HLB 200 + Strata-X-AW(100)/ Strata-X-CW(100)/ Isoelute ENV+(150) | 5 mL MeOH/ACN (2:8) + 6 mL 0.5% NH$_3$ in MeOH/ACN (2:8) + 4 mL 1.7% HCOOH in MeOH/ACN (2:8) | 22–112% | 0.1–58 | [54] |
| BTZ | wastewater | Oasis HLB 200 + Strata-X-AW(100)/ Strata-X-CW(100)/ Isoelute ENV+(150) + ENVI-Carb (200) | 6 mL EA/ MeOH/NH$_3$ + 3 mL EA/MeOH/ HCOOH | 4–24% | ND | [56] |
| BTZ, 5MBZ, CIBZ, DMBZ | river, groundwater | Oasis HLB 500 | 3 × 2 mL MeOH/DCM (1:1) | 71–95% | 0.8–1.3 | [57] |

Oasis series is produced by Waters, Strata series by Phenomenex, Bond Elut by Agilent, Poly-sery by Anpel, PLRP-s by Spark, Isoelute ENV+ by Biotage, Supel-Select and ENVI-Carb by Supelco.

### 5. Solid-Phase Extraction of Benzotriazoles from Solid and Other Samples

SPE can be also used for clean-up of solid sample extracts obtained from sediments, activated sludge, soil, plant and detergent matrices, as presented in Table 3. The original samples have been either simply dissolved [6,44], or extracted with liquid–solid extraction (LSE) with ultrasonication, accelerated solvent extraction (ASE [14]), pressurized hot water extraction (PHWE [75]) or microwave extraction (MAE [76]).

In addition to environmental samples, polar benzotriazoles have also been determined in other samples, presented in Table 4. The analysis of detergents and industrial fluids is important for investigations of the point of origin of these compounds in the environment, while an estimation of human and foetal intake of benzotriazoles can be made from the concentrations found in urine and amniotic fluid.

Because SPE is such a staple in analytical chemistry, it is rarely compared to other extraction techniques any more, but Asimakopoulos et al. [17] have compared extractions with two SPE cartridges (Strata-X-CW and Oasis HLB) to LLE with ACN/DCM for benzotriazoles in human urine. While recoveries were acceptable with both methods (92–127% SPE with HLB, 84–90% with LLE), the researchers noted high background contamination levels in LLE and also a co-elution of an isobaric component with BTZ, which led to higher LOD (1.17 ng/mL with LLE vs. 0.38 ng/mL with SPE). Suppression of ionization was noticed for both methods; but in real samples, lower concentrations of MBZ and DMBZ have been determined when using the LLE method than with SPE.

**Table 3.** SPE methods reported for benzotriazoles in solid samples (ND–no data).

| Compounds | Sample Matrices | Cartidge Type, mg | Elution Solvent | Recovery | LOQ (ng/g) | Reference |
|---|---|---|---|---|---|---|
| BTZ, 5MBZ | detergent | Oasis HLB 500 | 5 mL 3% MeOH in DCM | ND | ND | [6] |
| BTZ, 5MBZ | river sediments, sludge | LSE (MeOH) + Oasis HLB 500 | 6 mL 15% MeOH in EA | 70–226% | 0.22 | [43] |
| BTZ, 5MBZ, DMBZ, 1OHBZ | sludge, suspended particles | LSE (MeOH/H$_2$O) + Strata-X 200 | 2 × 5 mL MeOH/ACN (1:1) | | | [39] |
| BTZ, 1MBZ, 4MBZ, 5MBZ | detergent, sludge | Oasis HLB 200 | 10 mL EA | 80–93% | ND | [44] |
| BTZ, 4MBZ, 5MBZ, DMBZ, 1OHBZ, CIBZ | house dust | LSE (MeOH/H$_2$O) + Oasis MAX 60 | 5 mL MeOH | 71–108% | 0.5 | [15] |
| BTZ, 4MBZ, 5MBZ, DMBZ, CIBZ | sludge | PHWE + Oasis HLB 150 + Florisil 500 | 2 × 3 mL MeOH | 39–89% | 0.5 | [75] |
| BTZ, 5MBZ | detergent | Oasis HLB 500 | 2 × 5 mL H$_2$O (pH 2.9)/MeOH (95:5) | ND | ND | [22] |
| BTZ, 5MBZ | estuary sediments | ASE + Oasis HLB | 6 mL MeOH | ND | 1.5–1.8 | [14] |
| BTZ, 4MBZ, 5MBZ, DMBZ, CIBZ | sludge | LSE (MeOH/H$_2$O) + Strata-X 200 | 10 mL MeOH/ACN (1:1) | 51–77% | 118–1666 | [65] |
| BTZ | plants | LSE (MeOH/H$_2$O) + Oasis HLB 500 | 5 mL MeOH + 5 mL MTBE/MeOH (9:1) | 100.5% | ND | [11] |
| 5MBZ | soil | LSE (ACO/hexane) + Strata-X 100 | 10 mL EA | 92–110% | 0.002–0.019 | [45] |
| BTZ, MBZ, DMBZ, CIBZ | sludge | LSE (MeOH/H$_2$O) + Oasis MCX 60 | 6 mL MeOH + 4 mL ACN | ND | ND | [58] |
| BTZ, MBZ | sludge | LSE (MeOH) + Oasis HLB 200 | 5 mL 5% HCOOH in MeOH + 5 mL EA | 97–9% | 20 | [70] |
| BTZ | sludge | MAE + Oasis HLB 250 | 5 mL MeOH + 5 mL MeOH/MTBE (1:9) | ND | ND | [76] |

Oasis series is produced by Waters, Strata series by Phenomenex.

**Table 4.** SPE methods reported for benzotriazoles in human samples and industrial liquids (ND–no data).

| Compounds | Sample Matrices | Cartidge Type, mg | Elution Solvent | Recovery | LOQ (ng/L) | Reference |
|---|---|---|---|---|---|---|
| BTZ, 4MBZ, 5MBZ | detergents; anti-icing, de-icing fluid | Bond Elut PPL 200 | 2 mL ACN/MeOH (1:1) | 36–41% (anti-icing) | 6–15 | [52] |
| BTZ, 5MBZ, DMBZ, 1OHBZ | human urine | Oasis HLB 200 | 10 mL MeOH/ACN (1:1) | 93–117% | 0.2–5 | [17] |
| BTZ | mineral oil | Sep-pak Plus 500 | 5 mL H$_2$O/ACN (4:6) | 77% | ND | [36] |
| BTZ, MBZ, 1OHBZ, CIBZ | human urine, amniotic fluid | Oasis HLB 200 | 10 mL MeOH/ACN (1:1) | urine: 67–106%, amn.fl. 71–93% | 5–510 | [18] |

Oasis series and Sep-pak Plus are produced by Waters, Bond Elut by Agilent.

## 6. Conclusions

Although in recent years the trends toward miniaturization of extraction techniques and an overall greener approach have led to the development of various new analytical approaches, SPE remains the mainstay of environmental chemistry. It is one of the first choices for researchers when encountering organic pollutants and this stays true also for emerging contaminants such as benzotriazoles. In recent years, many methods have been developed for the determination of polar benzotriazoles, not only for aqueous but also for solid environmental, industrial and biological samples, as presented in this review. The general trends and advances in SPE seem to be focused on reducing the matrix effects and investigating various newer sorbent phases, which is reflected also in the study of benzotriazoles, where quite a few new studies were done with mixed-mode sorbents. Benzotriazoles are also often included in screening studies of environmental samples, where the trend of more and more analytes being determined in one analysis can be observed. The amount of publications found in this corner of environmental chemistry leads us to conclude that despite its long tradition, SPE has many uses for research left.

**Author Contributions:** Writing—original draft preparation: I.K. and H.P.; writing—review and editing: I.K. and H.P.

**Funding:** The authors would like to acknowledge the financial support of Slovenian Research Agency (Grant MR 38132 and research core funding No. P1-0153).

**Conflicts of Interest:** The authors declare no conflict of interest.

## Abbreviations

The following abbreviations are used in this manuscript:

| | |
|---|---|
| ACN | acetonitrile |
| ACO | acetone |
| DCM | dichloromethane |
| EA | ethyl acetate |
| GC | gas chromatography |
| LC | liquid chromatography |
| LLE | liquid–liquid extraction |
| LSE | liquid–solid extraction |
| LOD | limit of detection |
| LOQ | limit of quantification |
| MeOH | methanol |
| MS | mass spectrometry |
| MTBE | methyl *tert*-butyl ether |

## References

1.  Simpson, N.J.K. (Ed.) *Solid-Phase Extraction: Principles, Techniques, and Applications*; Marcel Dekker: New York, NY, USA, 2000.
2.  Reemtsma, T.; Miehe, U.; Duennbier, U.; Jekel, M. Polar pollutants in municipal wastewater and the water cycle: Occurrence and removal of benzotriazoles. *Water Res.* **2010**, *44*, 596–604. [CrossRef] [PubMed]
3.  Weiss, S.; Reemtsma, T. Determination of Benzotriazole Corrosion Inhibitors from Aqueous Environmental Samples by Liquid Chromatography-Electrospray Ionization-Tandem Mass Spectrometry. *Anal. Chem.* **2005**, *77*, 7415–7420. [CrossRef] [PubMed]
4.  Herrero, P.; Borrull, F.; Pocurull, E.; Marcé, R. Efficient tandem solid-phase extraction and liquid chromatography-triple quadrupole mass spectrometry method to determine polar benzotriazole, benzothiazole and benzenesulfonamide contaminants in environmental water samples. *J. Chromatogr. A* **2013**, *1309*, 22–32. [CrossRef] [PubMed]
5.  Kiss, A.; Fries, E. Occurrence of benzotriazoles in the rivers Main, Hengstbach, and Hegbach (Germany). *Environ. Sci. Pollut. Res.* **2009**, *16*, 702–710. [CrossRef] [PubMed]

6.    Janna, H.; Scrimshaw, M.D.; Williams, R.J.; Churchley, J.; Sumpter, J.P. From Dishwasher to Tap? Xenobiotic Substances Benzotriazole and Tolyltriazole in the Environment. *Environ. Sci. Technol.* **2011**, *45*, 3858–3864. [CrossRef] [PubMed]

7.    Liu, Y.S.; Ying, G.G.; Shareef, A.; Kookana, R.S. Simultaneous determination of benzotriazoles and ultraviolet filters in ground water, effluent and biosolid samples using gas chromatography–tandem mass spectrometry. *J. Chromatogr. A* **2011**, *1218*, 5328–5335. [CrossRef] [PubMed]

8.    Carpinteiro, I.; Abuin, B.; Ramil, M.; Rodríguez, I.; Cela, R. Simultaneous determination of benzotriazole and benzothiazole derivatives in aqueous matrices by mixed-mode solid-phase extraction followed by liquid chromatography–tandem mass spectrometry. *Anal. Bioanal. Chem.* **2012**, *402*, 2471–2478. [CrossRef] [PubMed]

9.    Nödler, K.; Voutsa, D.; Licha, T. Polar organic micropollutants in the coastal environment of different marine systems. *Mar. Pollut. Bull.* **2014**, *85*, 50–59. [CrossRef] [PubMed]

10.   Wang, L.; Zhang, J.; Sun, H.; Zhou, Q. Widespread Occurrence of Benzotriazoles and Benzothiazoles in Tap Water: Influencing Factors and Contribution to Human Exposure. *Environ. Sci. Technol.* **2016**, *50*, 2709–2717. [CrossRef] [PubMed]

11.   LeFevre, G.H.; Lipsky, A.; Hyland, K.C.; Blaine, A.C.; Higgins, C.P.; Luthy, R.G. Benzotriazole (BT) and BT plant metabolites in crops irrigated with recycled water. *Environ. Sci. Water Res. Technol.* **2017**, *3*, 213–223. [CrossRef]

12.   Xue, J.; Wan, Y.; Kannan, K. Occurrence of benzotriazoles (BTRs) in indoor air from Albany, New York, USA, and its implications for inhalation exposure. *Toxicol. Environ. Chem.* **2017**, *99*, 402–414. [CrossRef]

13.   Speltini, A.; Sturini, M.; Maraschi, F.; Porta, A.; Profumo, A. Fast low-pressurized microwave-assisted extraction of benzotriazole, benzothiazole and benezenesulfonamide compounds from soil samples. *Talanta* **2016**, *147*, 322–327. [CrossRef] [PubMed]

14.   Cantwell, M.G.; Sullivan, J.C.; Katz, D.R.; Burgess, R.M.; Bradford Hubeny, J.; King, J. Source determination of benzotriazoles in sediment cores from two urban estuaries on the Atlantic Coast of the United States. *Mar. Pollut. Bull.* **2015**, *101*, 208–218. [CrossRef] [PubMed]

15.   Wang, L.; Asimakopoulos, A.G.; Moon, H.B.; Nakata, H.; Kannan, K. Benzotriazole, Benzothiazole, and Benzophenone Compounds in Indoor Dust from the United States and East Asian Countries. *Environ. Sci. Technol.* **2013**, *47*, 4752–4759. [CrossRef] [PubMed]

16.   Luongo, G.; Avagyan, R.; Hongyu, R.; Östman, C. The washout effect during laundry on benzothiazole, benzotriazole, quinoline, and their derivatives in clothing textiles. *Environ. Sci. Pollut. Res.* **2016**, *23*, 2537–2548. [CrossRef] [PubMed]

17.   Asimakopoulos, A.G.; Bletsou, A.A.; Wu, Q.; Thomaidis, N.S.; Kannan, K. Determination of Benzotriazoles and Benzothiazoles in Human Urine by Liquid Chromatography-Tandem Mass Spectrometry. *Anal. Chem.* **2013**, *85*, 441–448. [CrossRef] [PubMed]

18.   Li, X.; Wang, L.; Asimakopoulos, A.G.; Sun, H.; Zhao, Z.; Zhang, J.; Zhang, L.; Wang, Q. Benzotriazoles and benzothiazoles in paired maternal urine and amniotic fluid samples from Tianjin, China. *Chemosphere* **2018**, *199*, 524–530. [CrossRef] [PubMed]

19.   Voutsa, D.; Hartmann, P.; Schaffner, C.; Giger, W. Benzotriazoles, Alkylphenols and Bisphenol A in Municipal Wastewaters and in the Glatt River, Switzerland. *Environ. Sci. Pollut. Res.* **2006**, *13*, 333–341. [CrossRef]

20.   Liu, Y.S.; Ying, G.G.; Shareef, A.; Kookana, R.S. Biodegradation of three selected benzotriazoles under aerobic and anaerobic conditions. *Water Res.* **2011**, *45*, 5005–5014. [CrossRef] [PubMed]

21.   van Leerdam, J.A.; Hogenboom, A.C.; van der Kooi, M.M.; de Voogt, P. Determination of polar 1H-benzotriazoles and benzothiazoles in water by solid-phase extraction and liquid chromatography LTQ FT Orbitrap mass spectrometry. *Int. J. Mass Spectrom.* **2009**, *282*, 99–107. [CrossRef]

22.   Alotaibi, M.D.; Patterson, B.M.; McKinley, A.J.; Reeder, A.Y.; Furness, A.J. Benzotriazole and 5-methylbenzotriazole in recycled water, surface water and dishwashing detergents from Perth, Western Australia: analytical method development and application. *Environ. Sci. Process. Impacts* **2015**, *17*, 448–457. [CrossRef] [PubMed]

23.   Deeb, A.A.; Schmidt, T.C. Tandem anion and cation exchange solid phase extraction for the enrichment of micropollutants and their transformation products from ozonation in a wastewater treatment plant. *Anal. Bioanal. Chem.* **2016**, *408*, 4219–4232. [CrossRef] [PubMed]

24. Bahnmüller, S.; Loi, C.H.; Linge, K.L.; Gunten, U.V.; Canonica, S. Degradation rates of benzotriazoles and benzothiazoles under UV-C irradiation and the advanced oxidation process UV/H$_2$O$_2$. *Water Res.* **2015**, *74*, 143–154. [CrossRef] [PubMed]

25. Seeland, A.; Oetken, M.; Kiss, A.; Fries, E.; Oehlmann, J. Acute and chronic toxicity of benzotriazoles to aquatic organisms. *Environ. Sci. Pollut. Res.* **2012**, *19*, 1781–1790. [CrossRef] [PubMed]

26. Harris, C.A.; Routledge, E.J.; Schaffner, C.; Brian, J.V.; Giger, W.; Sumpter, J.P. Benzotriazole is antiestrogenic in vitro but not in vivo. *Environ. Toxicol. Chem.* **2007**, *26*, 2367–2372. [CrossRef] [PubMed]

27. Pervova, M.G.; Kirichenko, V.E.; Saloutin, V.I. Determination of 1,2,3-benzotriazole in aqueous solutions and air by reaction-gas-liquid chromatography. *J. Anal. Chem.* **2010**, *65*, 276–279. [CrossRef]

28. Herrero, P.; Borrull, F.; Pocurull, E.; Marcé, R.M. A quick, easy, cheap, effective, rugged and safe extraction method followed by liquid chromatography-(Orbitrap) high resolution mass spectrometry to determine benzotriazole, benzothiazole and benzenesulfonamide derivates in sewage sludge. *J. Chromatogr. A* **2014**, *1339*, 34–41. [CrossRef] [PubMed]

29. Gilart, N.; Cormack, P.A.G.; Marcé, R.M.; Borrull, F.; Fontanals, N. Preparation of a polar monolithic coating for stir bar sorptive extraction of emerging contaminants from wastewaters. *J. Chromatogr. A* **2013**, *1295*, 42–47. [CrossRef] [PubMed]

30. Naccarato, A.; Gionfriddo, E.; Sindona, G.; Tagarelli, A. Simultaneous determination of benzothiazoles, benzotriazoles and benzosulfonamides by solid phase microextraction-gas chromatography-triple quadrupole mass spectrometry in environmental aqueous matrices and human urine. *J. Chromatogr. A* **2014**, *1338*, 164–173. [CrossRef] [PubMed]

31. Casado, J.; Rodríguez, I.; Ramil, M.; Cela, R. Polyethersulfone solid-phase microextraction followed by liquid chromatography quadrupole time-of-flight mass spectrometry for benzotriazoles determination in water samples. *J. Chromatogr. A* **2013**, *1299*, 40–47. [CrossRef] [PubMed]

32. Pena, M.T.; Vecino-Bello, X.; Casais, M.C.; Mejuto, M.C.; Cela, R. Optimization of a dispersive liquid–liquid microextraction method for the analysis of benzotriazoles and benzothiazoles in water samples. *Anal. Bioanal. Chem.* **2012**, *402*, 1679–1695. [CrossRef] [PubMed]

33. Casado, J.; Nescatelli, R.; Rodríguez, I.; Ramil, M.; Marini, F.; Cela, R. Determination of benzotriazoles in water samples by concurrent derivatization–dispersive liquid–liquid microextraction followed by gas chromatography–mass spectrometry. *J. Chromatogr. A* **2014**, *1336*, 1–9. [CrossRef] [PubMed]

34. Alotaibi, M.D.; McKinley, A.J.; Patterson, B.M.; Reeder, A.Y. Benzotriazoles in the Aquatic Environment: a Review of Their Occurrence, Toxicity, Degradation and Analysis. *Water Air Soil Pollut.* **2015**, *226*. [CrossRef]

35. Herrero, P.; Borrull, F.; Pocurull, E.; Marcé, R.M. An overview of analytical methods and occurrence of benzotriazoles, benzothiazoles and benzenesulfonamides in the environment. *Trends Anal. Chem.* **2014**, *62*, 46–55. [CrossRef]

36. Bruzzoniti, M.C.; Maina, R.; Tumiatti, V.; Sarzanini, C.; De Carlo, R.M. Simultaneous Determination of Passivator and Antioxidant Additives in Insulating Mineral Oils by High-Performance Liquid Chromatography. *J. Liq. Chromatogr. Relat. Technol.* **2015**, *38*, 15–19. [CrossRef]

37. Hogenboom, A.; van Leerdam, J.; de Voogt, P. Accurate mass screening and identification of emerging contaminants in environmental samples by liquid chromatography–hybrid linear ion trap Orbitrap mass spectrometry. *J. Chromatogr. A* **2009**, *1216*, 510–519. [CrossRef] [PubMed]

38. Salas, D.; Borrull, F.; Marcé, R.M.; Fontanals, N. Study of the retention of benzotriazoles, benzothiazoles and benzenesulfonamides in mixed-mode solid-phase extraction in environmental samples. *J. Chromatogr. A* **2016**, *1444*, 21–31. [CrossRef] [PubMed]

39. Asimakopoulos, A.G.; Ajibola, A.; Kannan, K.; Thomaidis, N.S. Occurrence and removal efficiencies of benzotriazoles and benzothiazoles in a wastewater treatment plant in Greece. *Sci. Total Environ.* **2013**, *452–453*, 163–171. [CrossRef] [PubMed]

40. Wolschke, H.; Xie, Z.; Möller, A.; Sturm, R.; Ebinghaus, R. Occurrence, distribution and fluxes of benzotriazoles along the German large river basins into the North Sea. *Water Res.* **2011**, *45*, 6259–6266. [CrossRef] [PubMed]

41. Loos, R.; Tavazzi, S.; Mariani, G.; Suurkuusk, G.; Paracchini, B.; Umlauf, G. Analysis of emerging organic contaminants in water, fish and suspended particulate matter (SPM) in the Joint Danube Survey using solid-phase extraction followed by UHPLC-MS-MS and GC–MS analysis. *Sci. Total Environ.* **2017**, *607–608*, 1201–1212. [CrossRef] [PubMed]

42. Molins-Delgado, D.; Távora, J.; Silvia Díaz-Cruz, M.; Barceló, D. UV filters and benzotriazoles in urban aquatic ecosystems: The footprint of daily use products. *Sci. Total Environ.* **2017**, *601–602*, 975–986. [CrossRef] [PubMed]

43. Zhang, Z.; Ren, N.; Li, Y.F.; Kunisue, T.; Gao, D.; Kannan, K. Determination of benzotriazole and benzophenone UV filters in sediment and sewage sludge. *Environ. Sci. Technol.* **2011**, *45*, 3909–3916. [CrossRef] [PubMed]

44. Spahr, S.; Huntscha, S.; Bolotin, J.; Maier, M.P.; Elsner, M.; Hollender, J.; Hofstetter, T.B. Compound-specific isotope analysis of benzotriazole and its derivatives. *Anal. Bioanal. Chem.* **2013**, *405*, 2843–2856. [CrossRef] [PubMed]

45. Hurtado, C.; Montano-Chávez, Y.N.; Domínguez, C.; Bayona, J.M. Degradation of Emerging Organic Contaminants in an Agricultural Soil: Decoupling Biotic and Abiotic Processes. *Water Air Soil Pollut.* **2017**, *228*, 243. [CrossRef]

46. Liu, Y.S.; Ying, G.G.; Shareef, A.; Kookana, R.S. Occurrence and removal of benzotriazoles and ultraviolet filters in a municipal wastewater treatment plant. *Environ. Pollut.* **2012**, *165*, 225–232. [CrossRef] [PubMed]

47. Jover, E.; Matamoros, V.; Bayona, J.M. Characterization of benzothiazoles, benzotriazoles and benzosulfonamides in aqueous matrixes by solid-phase extraction followed by comprehensive two-dimensional gas chromatography coupled to time-of-flight mass spectrometry. *J. Chromatogr. A* **2009**, *1216*, 4013–4019. [CrossRef] [PubMed]

48. Liu, Y.S.; Ying, G.G.; Shareef, A.; Kookana, R.S. Biodegradation of three selected benzotriazoles in aquifer materials under aerobic and anaerobic conditions. *J. Contam. Hydrol.* **2013**, *151*, 131–139. [CrossRef] [PubMed]

49. Sulej, A.M.; Polkowska, Ż.; Astel, A.; Namieśnik, J. Analytical procedures for the determination of fuel combustion products, anti-corrosive compounds, and de-icing compounds in airport runoff water samples. *Talanta* **2013**, *117*, 158–167. [CrossRef] [PubMed]

50. Koroša, A.; Auersperger, P.; Mali, N. Determination of micro-organic contaminants in groundwater (Maribor, Slovenia). *Sci. Total Environ.* **2016**, *571*, 1419–1431. [CrossRef] [PubMed]

51. Domínguez, C.; Reyes-Contreras, C.; Bayona, J.M. Determination of benzothiazoles and benzotriazoles by using ionic liquid stationary phases in gas chromatography mass spectrometry. Application to their characterization in wastewaters. *J. Chromatogr. A* **2012**, *1230*, 117–122. [CrossRef] [PubMed]

52. Kiss, A.; Fries, E. Seasonal source influence on river mass flows of benzotriazoles. *J. Environ. Monit.* **2012**, *14*, 697. [CrossRef] [PubMed]

53. Loos, R.; Tavazzi, S.; Paracchini, B.; Canuti, E.; Weissteiner, C. Analysis of polar organic contaminants in surface water of the northern Adriatic Sea by solid-phase extraction followed by ultrahigh-pressure liquid chromatography–QTRAP<Superscript>®</Superscript> MS using a hybrid triple-quadrupole linear ion trap instrument. *Anal. Bioanal. Chem.* **2013**, *405*, 5875–5885. [CrossRef] [PubMed]

54. Osorio, V.; Schriks, M.; Vughs, D.; de Voogt, P.; Kolkman, A. A novel sample preparation procedure for effect-directed analysis of micro-contaminants of emerging concern in surface waters. *Talanta* **2018**, *186*, 527–537. [CrossRef] [PubMed]

55. Thellmann, P.; Köhler, H.R.; Rößler, A.; Scheurer, M.; Schwarz, S.; Vogel, H.J.; Triebskorn, R. Fish embryo tests with Danio rerio as a tool to evaluate surface water and sediment quality in rivers influenced by wastewater treatment plants using different treatment technologies. *Environ. Sci. Pollut. Res.* **2015**, *22*, 16405–16416. [CrossRef] [PubMed]

56. Xing, Y.; Yu, Y.; Men, Y. Emerging investigators series: Occurrence and fate of emerging organic contaminants in wastewater treatment plants with an enhanced nitrification step. *Environ. Sci. Water Res. Technol.* **2018**. [CrossRef]

57. Yang, Y.Y.; Zhao, J.L.; Liu, Y.S.; Liu, W.R.; Zhang, Q.Q.; Yao, L.; Hu, L.X.; Zhang, J.N.; Jiang, Y.X.; Ying, G.G. Pharmaceuticals and personal care products (PPCPs) and artificial sweeteners (ASs) in surface and ground waters and their application as indication of wastewater contamination. *Sci. Total Environ.* **2018**, *616–617*, 816–823. [CrossRef] [PubMed]

58. Karthikraj, R.; Kannan, K. Mass loading and removal of benzotriazoles, benzothiazoles, benzophenones, and bisphenols in Indian sewage treatment plants. *Chemosphere* **2017**, *181*, 216–223. [CrossRef] [PubMed]

59. Huntscha, S.; Singer, H.P.; McArdell, C.S.; Frank, C.E.; Hollender, J. Multiresidue analysis of 88 polar organic micropollutants in ground, surface and wastewater using online mixed-bed multilayer solid-phase extraction coupled to high performance liquid chromatography–tandem mass spectrometry. *J. Chromatogr. A* **2012**, *1268*, 74–83. [CrossRef] [PubMed]

60. Busetti, F.; Ruff, M.; Linge, K.L. Target screening of chemicals of concern in recycled water. *Environ. Sci. Water Res. Technol.* **2015**, *1*, 659–667. [CrossRef]

61. Bourgin, M.; Beck, B.; Boehler, M.; Borowska, E.; Fleiner, J.; Salhi, E.; Teichler, R.; von Gunten, U.; Siegrist, H.; McArdell, C.S. Evaluation of a full-scale wastewater treatment plant upgraded with ozonation and biological post-treatments: Abatement of micropollutants, formation of transformation products and oxidation by-products. *Water Res.* **2018**, *129*, 486–498. [CrossRef] [PubMed]

62. Loos, R.; Locoro, G.; Contini, S. Occurrence of polar organic contaminants in the dissolved water phase of the Danube River and its major tributaries using SPE-LC-MS2 analysis. *Water Res.* **2010**, *44*, 2325–2335. [CrossRef] [PubMed]

63. Loos, R.; Locoro, G.; Comero, S.; Contini, S.; Schwesig, D.; Werres, F.; Balsaa, P.; Gans, O.; Weiss, S.; Blaha, L.; et al. Pan-European survey on the occurrence of selected polar organic persistent pollutants in ground water. *Water Res.* **2010**, *44*, 4115–4126. [CrossRef] [PubMed]

64. Loos, R.; Carvalho, R.; António, D.C.; Comero, S.; Locoro, G.; Tavazzi, S.; Paracchini, B.; Ghiani, M.; Lettieri, T.; Blaha, L.; et al. EU-wide monitoring survey on emerging polar organic contaminants in wastewater treatment plant effluents. *Water Res.* **2013**, *47*, 6475–6487. [CrossRef] [PubMed]

65. Mazioti, A.A.; Stasinakis, A.S.; Gatidou, G.; Thomaidis, N.S.; Andersen, H.R. Sorption and biodegradation of selected benzotriazoles and hydroxybenzothiazole in activated sludge and estimation of their fate during wastewater treatment. *Chemosphere* **2015**, *131*, 117–123. [CrossRef] [PubMed]

66. Gatidou, G.; Oursouzidou, M.; Stefanatou, A.; Stasinakis, A.S. Removal mechanisms of benzotriazoles in duckweed Lemna minor wastewater treatment systems. *Sci. Total Environ.* **2017**, *596–597*, 12–17. [CrossRef] [PubMed]

67. Molins-Delgado, D.; Silvia Díaz-Cruz, M.; Barceló, D. Removal of polar UV stabilizers in biological wastewater treatments and ecotoxicological implications. *Chemosphere* **2015**, *119*, S51–S57. [CrossRef] [PubMed]

68. Serra-Roig, M.P.; Jurado, A.; Díaz-Cruz, M.S.; Vázquez-Suñé, E.; Pujades, E.; Barceló, D. Occurrence, fate and risk assessment of personal care products in river–groundwater interface. *Sci. Total Environ.* **2016**, *568*, 829–837. [CrossRef] [PubMed]

69. Lu, J.; Mao, H.; Li, H.; Wang, Q.; Yang, Z. Occurrence of and human exposure to parabens, benzophenones, benzotriazoles, triclosan and triclocarban in outdoor swimming pool water in Changsha, China. *Sci. Total Environ.* **2017**, *605–606*, 1064–1069. [CrossRef] [PubMed]

70. Östman, M.; Lindberg, R.H.; Fick, J.; Björn, E.; Tysklind, M. Screening of biocides, metals and antibiotics in Swedish sewage sludge and wastewater. *Water Res.* **2017**, *115*, 318–328. [CrossRef] [PubMed]

71. Chung, K.H.Y.; Lin, Y.C.; Lin, A.Y.C. The persistence and photostabilizing characteristics of benzotriazole and 5-methyl-1H-benzotriazole reduce the photochemical behavior of common photosensitizers and organic compounds in aqueous environments. *Environ. Sci. Pollut. Res.* **2018**, *25*, 5911–5920. [CrossRef] [PubMed]

72. Lai, W.W.P.; Lin, Y.C.; Wang, Y.H.; Guo, Y.L.; Lin, A.Y.C. Occurrence of Emerging Contaminants in Aquaculture Waters: Cross-Contamination between Aquaculture Systems and Surrounding Waters. *Water Air Soil Pollut.* **2018**, *229*, 249. [CrossRef]

73. Milić, N.; Milanović, M.; Radonić, J.; Sekulić, M.T.; Mandić, A.; Orčić, D.; Mišan, A.; Milovanović, I.; Letić, N.G.; Miloradov, M.V. The occurrence of selected xenobiotics in the Danube river via LC-MS/MS. *Environ. Sci. Pollut. Res.* **2018**, *25*, 11074–11083. [CrossRef] [PubMed]

74. Lu, J.; Li, H.; Luo, Z.; Lin, H.; Yang, Z. Occurrence, distribution, and environmental risk of four categories of personal care products in the Xiangjiang River, China. *Environ. Sci. Pollut. Res.* **2018**, 1–11. [CrossRef] [PubMed]

75. Herrero, P.; Borrull, F.; Marcé, R.M.; Pocurull, E. A pressurised hot water extraction and liquid chromatography–high resolution mass spectrometry method to determine polar benzotriazole, benzothiazole and benzenesulfonamide derivates in sewage sludge. *J. Chromatogr. A* **2014**, *1355*, 53–60. [CrossRef] [PubMed]

76. Lakshminarasimman, N.; Quiñones, O.; Vanderford, B.J.; Campo-Moreno, P.; Dickenson, E.V.; McAvoy, D.C. Biotransformation and sorption of trace organic compounds in biological nutrient removal treatment systems. *Sci. Total Environ.* **2018**, *640–641*, 62–72. [CrossRef] [PubMed]

*molecules*

MDPI

*Article*

# Melamine Sponge Functionalized with Urea-Formaldehyde Co-Oligomers as a Sorbent for the Solid-Phase Extraction of Hydrophobic Analytes

María Teresa García-Valverde [1], Theodoros Chatzimitakos [2], Rafael Lucena [1], Soledad Cárdenas [1] and Constantine D. Stalikas [2,*]

[1] Grupo FQM-215, Departamento de Química Analítica, Instituto Universitario de Investigación en Química Fina y Nanoquímica (IUIQFN), Universidad de Córdoba, Campus de Rabanales, Edificio Marie Curie, E-14071 Córdoba, Spain; q72gavam@uco.es (M.T.G.-V.); rafael.lucena@uco.es (R.L.); scardenas@uco.es (S.C.)
[2] Laboratory of Analytical Chemistry, Department of Chemistry, University of Ioannina, 45110 Ioannina, Greece; chatzimitakos@outlook.com
* Correspondence: cstalika@cc.uoi.gr; Tel.: +30-265-100-8414

Academic Editor: Victoria F. Samanidou
Received: 15 September 2018; Accepted: 7 October 2018; Published: 10 October 2018

**Abstract:** A new procedure for the functionalization of melamine sponge (MeS) with urea-formaldehyde (UF) co-oligomers is put forward. The procedure differs from the typical synthesis of the UF co-polymer, as it employs a base-catalyzed condensation step at certain concentrations of urea and formaldehyde. The produced melamine-urea-formaldehyde (MUF) sponge cubes are hydrophobic, despite the presence of hydrophilic groups in the oligomers. The MUF sponge developed herein is used as a sorbent for the solid-phase extraction of 10 analytes, from 6 different classes (i.e., non-steroidal anti-inflammatory drugs, benzophenones, parabens, phenols, pesticides and musks) and an analytical method is developed for their liquid chromatographic separation and detection. Low limits of quantification (0.03 and 1.0 $\mu g\ L^{-1}$), wide linear ranges and excellent recoveries (92–100%) are some of the benefits of the proposed procedure. The study of the synthesis conditions of MUF cubes reveals that by altering them the hydrophilic/lipophilic balance of the MUF cubes can be tuned, hinting towards a strong potential for many other applications.

**Keywords:** solid-phase extraction; melamine sponge; urea-formaldehyde co-oligomers; HPLC-DAD

## 1. Introduction

Sample pretreatment is an essential part of any analytical method, whatever technique is subsequently applied, as it has major effect on the quality of the results obtained. Sample pretreatment methods are constantly being developed to respond to the large variety of sample sources and compositions [1]. Solid-phase extraction (SPE) is a highly useful sample preparation-pretreatment technique. Many of the problems associated with liquid/liquid extraction, such as incomplete phase separations, less-than-quantitative recoveries and disposal of large quantities of organic solvents can be prevented with SPE. Moreover, SPE helps meet the sample preparation challenges, with respect to analyte pre-concentration and sample clean-up [2–4].

In recent years, interconnected microporous sorbent materials with favorable (super)hydrophobic surface properties and excellent sorption capacity, have attracted significant interest for potential applications. A water droplet on such a sorbent surface can roll off, even at inclinations of only a few degrees, while still retaining oil contaminants encountered on its way into its matrixes [5,6]. Motivated by these properties, many efforts have been devoted to developing novel and advanced 3D porous materials, which possess hydrophobic surface properties and interconnected macro-porous structures, with excellent sorption performance, for the separation and recovery of organic pollutants from water [7–9]. It may be mentioned that such efforts in developing functional architectural materials have significantly improved the sorption performance towards various organic solvents. Melamine sponge (MeS) is a three-dimensional, low-density, foam-like material made of a formaldehyde-melamine-sodium bisulfite copolymer. It has an open-hole structure, high porosity (> 99%), excellent wettability and negligible cost to obtain [10,11] Because of the presence of functional groups, MeS is amenable to functionalization in order to tune its hydrophilicity or render it hydrophobic [12–14]. Such modifications have enabled the application of MeS-based hydrophobic materials to oil-absorption and oil/water separation [15–17].

Recently, we proposed for the first time, two modifications of MeS: first, with graphene in a one-step, fast procedure and we utilized it for the microextraction of sulfonamides from food and environmental matrixes [18] secondly, with copper sheets as a decorated material for the same purpose, capitalizing on the affinity of sulfonamides for copper, the rapid mass transport and on the low back-pressure due to their macropore and throughpore structure [19].

Melamine-urea-formaldehyde (MUF) resins are used in the manufacturing of water-resistant particleboards, papers and laminates. The chemistry of melamine reacting with formaldehyde and urea-formaldehyde (UF) is well known [20]. A molecularly imprinted monolithic resin has, also, been synthesized from MUF via an in-situ polymerization [21], which showed good water compatibility and excellent molecular recognition of plant growth regulators, when it was applied as an SPE sorbent. Melamine-urea-formaldehyde introduces abundant hydrophilic groups (such as hydroxyl, carboxyl, imino and amino groups) into the polymer and improves its compatibility with aqueous solutions and rapid mass transport.

The purpose of the study herein is to develop a novel, three-dimensional, low-density sponge-based hydrophobic material by properly developing formaldehyde-urea co-oligomers on the surface of a melamine sponge. This is achieved by carefully tuning the hydrophilic/lipophilic balance in favor of the latter. Melamine provides the backbone for the development of the oligomers. The modified melamine sponges are then used as an SPE material for the extraction of hydrophobic molecules belonging to six different classes of analytes.

## 2. Results and Discussion

### 2.1. Synthesis Optimization

The synthesis consisted of heating a urea-formaldehyde solution at alkaline pH for the formation of UF oligomers (*vide infra*) and their attachment to melamine to form the modified 3D material. Typically, UF polymer synthesis has been reported to consist of two steps: first, polymerization under alkaline conditions and secondly, condensation under acidic conditions [22]. However, this procedure leads to a highly hydrophilic polymer, owing to the occurrence of many hydrophilic groups (such as hydroxyl, carboxyl, imino and amino groups). In order to obtain a more hydrophobic product, in the present study, we employed a base-catalyzed UF condensation step, during which the formation of the products was kinetically and thermodynamically controlled. Under these conditions, the typical resin structure was not generated, while oligomers were primarily formed [23]. The MUF sponge synthesis was optimized so as to achieve maximum adsorption of the analytes. The ratio of formaldehyde: urea and the concentrations of both reagents were optimized with the purpose of ensuring efficient functionalization of MeS. Moreover, the number of sponges that can be produced in a single functionalization batch was studied. In all cases, the criterion used for the evaluation of the produced MUF cubes was the total extraction yield from an aqueous solution containing a mixture of analytes (50 µg L$^{-1}$ each). A complete list of the analytes, their structure and some physicochemical characteristics are given in Table 1.

**Table 1.** Structure and physicochemical properties of the studied analytes.

| Compound | Category | Structure | Log P | pK$_a$ | Quantification Wavelength | Retention Time (min) |
|---|---|---|---|---|---|---|
| Fenbufen (FEN) | Non-steroidal anti-inflammatory drugs | | 3.13 | 4.22 | 285 | 7.2 |
| Flurbiprofen (FLU) | | | 4.11 | 4.42 | 254 | 17.2 |
| Benzophenone-8 (BP8) | Benzophenones | | 3.93 | 7.11 | 285 | 12.0 |
| Butylparaben (BPB) | Parabens | | 3.46 | 8.47 | 254 | 9.8 |
| Cumylphenol (CUM) | Phenols | | 4.17 | 10.0 | 275 | 15.3 |
| 4-octylphenol (4-OP) | | | 5.66 | 10.15 | 275 | 24.8 |
| Chlorpyrifos (CLP) | Pesticides | | 4.77 | - | 285 | 18.4 |
| Trifluralin (TRIF) | | | 5.41 | - | 285 | 20.6 |
| Deltamethrin (DEL) | | | 6.20 | 10.62 | 275 | 31.1 |
| Tonalide (TON) | Musks | | 5.70 | - | 254 | 27.2 |

### 2.1.1. Effect of Formaldehyde-to-Urea Ratio

In order to study the formaldehyde-to-urea ratio (F/U), the amount of urea in the mixture was held constant at 0.88 mmol and formaldehyde was varied to probe its effect on the resulting MUF cubes. To this end, different products were obtained using F/U molar ratios of 1.0, 1.4, 1.6 and 2.0. Extraction of the analytes was carried out as explained in Section 3.4. According to the results (data not shown), the MUF cubes prepared at molar ratios of 1.0, 1.4 and 1.6 had similar extraction behavior for all analytes (the variations were lower than 7%). When a molar ratio of 2.0 was used, the total extraction yield was decreased by nearly 10%. This can be attributed to the prevalence of ether linkages (–CH$_2$–O–CH$_2$–) over methylene (–CH$_2$–) ones, at this ratio [23], which was unsuitable for producing MUF cubes with reasonable extraction yields for the target compounds. Ether bridges endow oligomers with hydrophilicity, compared to methyl analogues, which are undesirable in our case, since target compounds are adsorbed mainly via hydrophobic interactions (*vide infra*). Among the three other molar ratios with similar extraction behavior, that of 1.0 was finally opted for, because reproducible formation of a linear chain UF oligomer was favored in contrast to the branched UF oligomer structures produced at higher ratios [22,23].

### 2.1.2. Effect of Concentrations of Urea and Formaldehyde

To maximize the extraction potential of the MUF cubes, the concentrations of urea and formaldehyde were studied in depth. Different MUF cubes were synthesized using 0.44, 0.88, 1.76 and 3.52 mmol of each of the reagents. The results are summarized in Table S1 (Supplementary Materials). It was observed that as the amount of the reagents increased, the extraction yield slightly decreased for all analytes (from 0.44 to 3.52 mmol the adsorption decreased by nearly 10%). Moreover, it was observed that the hydrophobicity of the MUF cubes slightly decreased as the amount of reagents increased. To verify this finding, MUF cubes were prepared using 13.2 mmol of each reagent (30 times higher than the lower tested amount). It was found that these cubes were highly hydrophilic and their sorptive behavior was quite different from that of the proposed MUF cubes. Specifically, the hydrophilic MUF cubes were able to adsorb low percentages of FEN, FLU, BPB and BP8 (<20%) while the rest of the analytes were not adsorbed at all. This is probably due to the ionizable nature of the adsorbed compounds, which can interact with the carboxyl and amino groups of the UF oligomers, albeit to a lesser extent than with MUF cubes prepared using the proposed low reagent concentrations. High concentrations of the reagents favor the formation of branched structures, that contain fewer free surface groups and result in lower extraction yields [22,23]. This, in turn, bespeaks that all analytes are adsorbed primarily via hydrophobic interactions.

As mentioned above, the products of the base-catalyzed UF condensation step are kinetically and thermodynamically controlled. The formation of methylene ether bridges (a kinetically favorable process) is preferred to that of methylene bridges (a thermodynamically favorable process) [22,23]. This signifies that a hydrophilic/lipophilic balance of the oligomers can be maintained, which affects directly the prepared MUF cubes and depends on the quantities and ratios of the reagents used. When the amount of each reagent was lower than 0.44 mmol, the extraction yield of the MUF cubes was decreased, probably, as a result of the lower amount of UF oligomers attached to the MeS. As an optimum, the amount of each reagent was selected to be 0.44 mmol, satisfying both the best performance and the minimum reagent amount.

### 2.1.3. Number of Synthesized Sponge Cubes per Batch

Until now, single MUFs were prepared to examine the aforementioned conditions. However, the production of more than one cube per batch is a great advantage in terms of time and cost. For this reason, five MUF cubes were synthesized, simultaneously, in the same batch, by increasing proportionally the amounts of the reagents in the reaction mixture (i.e., from 0.44 mmol in 20 mL water per sponge to a total of 2.2 mmol in 100 mL of water). The extraction performance of the MUF cubes originating from a big batch was compared with that from the single-batch MUF cubes. The results revealed that no differences were recorded when bulk synthesis was carried out with more than one sponge at a time. Additionally, we examined the reproducibility of the extraction yield using MUF cubes from big batch syntheses, conducted on different days. In this case, also, the differences between various (big) batches from three consecutive days were insignificant (<6%). Therefore, five cubes were selected to be simultaneously modified according to the procedure.

### 2.2. Characterization of MUF Sponges

Figure S1 (Supplementary Materials) shows the SEM micrographs of raw MeS and MUF sponges. It can be seen that after functionalization, the fibrous 3D structure is maintained. The fact that bulk structures of UF are not observed strengthens the notion that using the base-catalyzed condensation step, oligomers are formed. Moreover, the typical UF resin structure was not observed. The porous framework of the cubes seems to shrink compared to the bare melamine analogues. This is verified by the reduction in cube size to nearly $0.9 \times 0.9 \times 0.9$ cm (from $1 \times 1 \times 1$ cm).

The FTIR spectra of pristine MeS, MUF cubes and UF resin can be seen in Figure 1. It is evident that after functionalization the IR spectrum of MUF cubes has some marked differences compared to that of MeS, evidencing the functionalization of MeS with UF, which differentiate it from typical UF resin. In contrast, the similarities suggest the successful functionalization of MeS with UF. The peaks at 1000, 1050 and 1250 cm$^{-1}$ (ether C–O stretch) do not appear in the spectrum of MUF, signifying that ether groups are not present, while the peak around 1450 cm$^{-1}$ is due to –CH$_2$ bending. The spectrum of UF polymer shows a large peak around 1650 cm$^{-1}$, due to amide group absorption. In the case of MeS and MUF cubes, the absorption bands around the above area are much weaker compared to those of UF polymer, suggesting that the amide groups in MUF cubes are limited. This supports the hydrophobic character of MUF cubes and strengthens our assertion that hydrophobic oligomers are formed under the stated synthesis conditions.

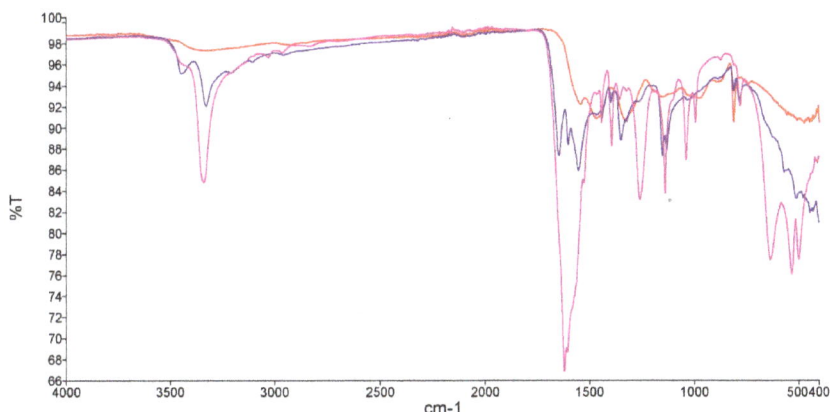

**Figure 1.** FTIR spectra of melamine sponge (MeS) (red line), urea-formaldehyde (UF) polymer (purple line) and melamine-urea-formaldehyde (MUF) cubes (blue line).

The wettability of the material was evaluated with the contact angle, which is a parameter that indicates the degree of wetting when a solid and liquid interact. Figure 2A shows a MUF cube, for which a red-colored (for visualization purposes) water drop was placed on the surface. The drop on the MUF sponge had an almost spherical shape with a contact angle of around 120°, demonstrating the hydrophobicity of the material. Moreover, when a MUF cube was placed in a glass beaker filled with water, the cube floated on the surface of the water, in contrast to bare MeS, which submerged easily to the bottom (Figure 2B). When MUF cubes were immersed in the water using metal forceps, the surface of the cube was like a silver mirror, consisting of many air bubbles, which is also characteristic of the hydrophobicity of the material (Figure 2C) [18,19].

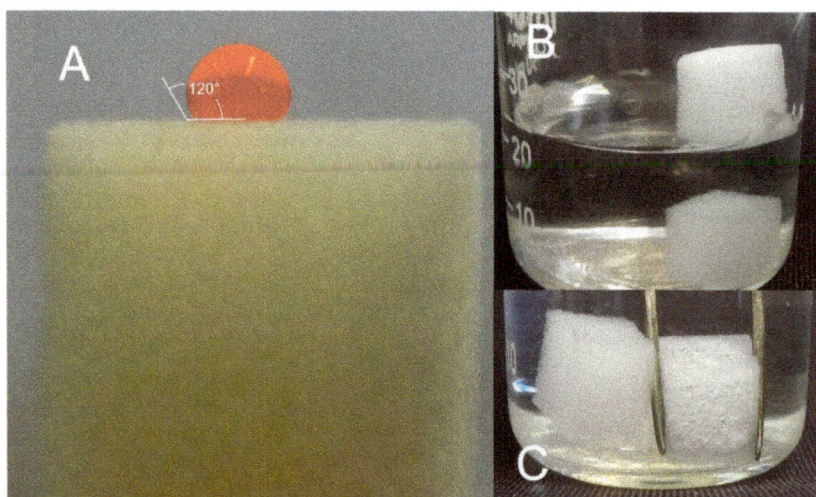

**Figure 2.** (**A**) A red colored water drop on the surface of a MUF cube and the approximate contact angle, (**B**) a MUF cube (upper) and a MeS (lower) in a glass beaker filled with water and (**C**) a MUF cube (right) immersed in water using forceps next to bare MeS (left).

## 2.3. Optimization of the Proposed Procedure

### 2.3.1. Effect of Sample pH and Ionic Strength on Extraction Yield

Due to the variety of classes of analytes employed in this study, a screening experiment in a broad pH range (i.e., 2.0–9.5) was carried out in order to elucidate the effect of pH on the extraction performance. The performance was improved at pH 4.5 for FEN, FLU, BPB, BP8, and CUM, while at pH 9.5 these analytes were hardly adsorbed (<5%). This is due to the ionizable nature of these compounds, hinting that their adsorption is not because of purely hydrophobic interactions. With regard to the rest of analytes, the variations in their extraction throughout the pH range of 2.0–7.0 where negligible, while a pronounced decrease by nearly 40% was noticed at pH 9.5. Figure 3 depicts the total extraction yield at different pH values, showing a favorable behavior at pH 4.5. As mentioned above, the adsorption of the analytes was achieved mainly via hydrophobic interactions and to a lesser extent (for some analytes) via other kind of interactions. For the subsequent experiments, the pH of the aqueous samples was adjusted to 4.5.

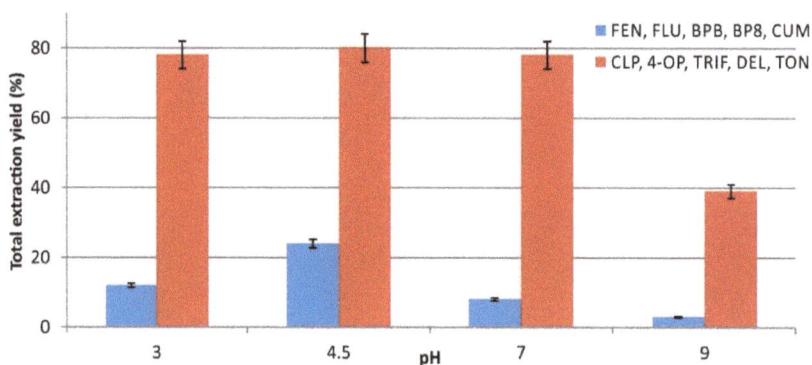

**Figure 3.** Effect of sample pH on the total extraction yield of the examined analytes, using the MUF cubes.

The effect of ionic strength on the extraction performance was evaluated using sodium chloride as model electrolyte. The concentrations tested were 0.25%, 0.5% and 1.0% $w/v$. The results showed that as the concentration of sodium chloride increases the extraction yield decreases. The addition of 0.25% and 0.5% $w/v$ of salt reduced the total extraction yield from 29% to 18.5%, while 1.0% $w/v$ of salt, further reduced adsorption to 13%. Although, commonly, salting out enhances the extraction yield of hydrophobic compounds, this was not the case in our method. This may be either due to the reduction of the diffusion rate of analytes, which, in general, is more pronounced for less polar analytes, or to the fact that increased salt concentrations cause an increase to the strong degree of sorption of the hydrophobic components, hindering their extraction [24,25]. Therefore, the deliberate increase of ionic strength of a sample prior to analysis is not beneficial.

### 2.3.2. Effect of Adsorbent Quantity and Sample Volume

The sorbent amount was studied by increasing the number of standard MUF cubes placed in the SPE cartridge, under constant flow rate (ca. 0.6 mL min$^{-1}$) with a sample volume of 25 mL. For this reason, one, two and three sponges were used simultaneously for the extraction of the analytes. At this step, elution was conducted by increasing proportionally the amount of the solvent used (500 µL methanol per cube), so that the observed differences were attributed solely to the extraction step. The performance of the extraction was improved as the number of sponges increased. Almost 15% and 25% increases were achieved with the second and the third MUF cubes, respectively, due to the marked increase in the available surface and sorbent functional groups of MUF. Adding a fourth MUF cube in the extraction unit was not found to further increase the extraction yield of the proposed procedure. Thus, three cubes were thereafter employed in order to achieve optimum extraction.

Regarding sample volume, experiments were carried out with different volumes spiked either with the same concentration of analytes, or with the same constant quantity of them. In both cases, the tested sample volumes were 10, 25, 50 and 75 mL. In the case of constant spiked quantity, the results showed a trivial decrease in the extraction efficiency of the target analytes (<7%) when sample volume was increased from 10 to 50 mL. In contrast, the extraction efficiency was lower by ~15% when the sample volume was 75 mL. In the case of different sample volumes, by keeping constant the concentration of analytes, the extraction efficiency was the same for 10 and 25 mL. A decrease of approximately 20% and 35% was noticed when the sample volume rose from 50 to 75 mL (see Figure S2 of Supplementary Materials). According to the above results, a sample volume of 25 mL was selected as the optimum, so as to attain significant enrichment factors of the analytes, with the proposed procedure.

### 2.3.3. Elution Conditions

Methanol and acetonitrile were tested as eluents for the proposed sample preparation method and the analytes of concern. Using methanol, 40% of the extracted analytes was eluted while acetonitrile was able to elute approximately 30%. As the extraction was found to be pH-dependent, we examined the presence of formic acid or ammonia in the eluent (in both cases 2% $v/v$ was added). When formic acid was added no significant differences were noticed for any analyte. However, when ammonia was added, elution yield sharply decreased by nearly 20% for FEN and FLU; for the rest of analytes, a trivial reduction of ~5–8% was noticed. Therefore, neither formic acid nor ammonia was finally added to methanol, which was selected as the optimum organic solvent.

As only 40% of the analytes were extracted under the above conditions, we examined the possibility of increasing this percentage by studying the elution volume. Preliminary experiments showed that an elution volume of less than 1 mL was not efficient when three sponges were used for the extraction. Therefore, 1 and 2 mL of methanol were subsequently examined. The elution yield was almost double when 2 mL of methanol were used (it was overall ~80%). Still, the analytes were not eluted completely from the sponges. Therefore, we conducted a second elution step with additional 2 mL of methanol following the first elution and we found that the analytes were totally recovered. In order to minimize the elution volume, the 2 mL of methanol already used for the first elution was passed, again, through the column and it was found that the extraction yield increased by 10%. A third elution step was conducted with the same eluent and the analytes were found to be removed completely from the cubes. Instead of using a total of 4 mL of methanol, we ended up using only 2 mL, reducing the solvent and the amount of time needed to evaporate the eluent using a nitrogen stream (from ~50 to ~35 min).

*2.4. Method Validation*

Under the aforementioned optimum conditions, the analytical figures of merit of the proposed procedure were studied. The analytical characteristics can be seen in detail, in Table 2. Calibration plots were drawn for each analyte, with the lowest concentrations as the limits of quantification (*LOQ*). Linear responses were noticed for all analytes up to a concentration of 100 µg L$^{-1}$. The coefficients of determination ($R^2$), in all cases, were higher than 0.9990, bespeaking satisfactory linearities. The limits of detection were between 0.01 and 0.33 µg L$^{-1}$. Enrichment factors (*EF*) and extraction percentages (*E*%) were calculated using the following equations:

$$EF = \frac{C_{el}}{C_{aq}} \tag{1}$$

$$E\% = \frac{C_{el}}{C_{el} + C_{aq}} \tag{2}$$

where, $C_{el}$ and $C_{aq}$ are the analyte concentrations in the eluent and the aqueous sample, respectively [26]. The values were found to be between 22% and 30% and above 96%, respectively, for all tested compounds. The precision of the method was calculated as the within-day and between-day relative standard deviation (*RSD*) by analyzing five different samples within the same day and three different samples on five consecutive days, respectively. The samples were spiked with the lowest concentrations employed for drawing the calibration plots. Within-day and between-day *RSD* were found to be between 5.6–7.3% and 6.1–8.4%, respectively.

*2.5. Analysis of Real Samples*

In order to examine the applicability of the proposed procedure, relative recoveries were calculated by analyzing lake water samples spiked with the analytes at two concentration levels, i.e., 3 and 10 times the *LOQ*. Typical chromatograms of a blank and a spiked lake water sample with a mixture of analytes (50 µg L$^{-1}$ each) are given in Figure S3 of the Supplementary Materials. The recoveries (summarized in Table 2) were in the ranges of 92–98% and 96–100% for the low and high tested concentrations, respectively. These values, along with *RSDs* and *LOQs* are comparable with those reported by other works, signifying that the proposed procedure is an excellent alternative method. A complete comparison with other methods can be seen in Table 3.

Table 2. Analytical figures of merit of the developed solid-phase extraction procedure, using MUF cubes [a].

| Compound | Linear Equation | Coefficient of Determination ($R^2$) | LOD ($\mu g\ L^{-1}$) | EF | E% | RSD (%) | | Relative Recoveries (%) | |
| --- | --- | --- | --- | --- | --- | --- | --- | --- | --- |
| | | | | | | Within-Day (n = 5) | Between-Day (n = 3 × 5) | 3 × LOQ | 10 × LOQ |
| Fenbufen | y = 31,481x + 1117 | 0.9998 | 0.02 | 22 | 96 | 7.0 | 7.4 | 94 | 96 |
| Butylparaben | y = 30,462x + 642 | 0.9993 | 0.01 | 23 | 96 | 6.7 | 7.4 | 95 | 98 |
| Benzophenone-8 | y = 30,121x + 832 | 0.9992 | 0.02 | 27 | 96 | 7.1 | 7.8 | 96 | 99 |
| Cumylphenol | y = 30,432x + 667 | 0.9991 | 0.02 | 23 | 96 | 6.4 | 7.1 | 95 | 98 |
| Flurbiprofen | y = 29,010x + 1844 | 0.9995 | 0.02 | 25 | 96 | 6.7 | 8.4 | 93 | 97 |
| Chlorpyrifos | y = 7498x + 1091 | 0.9993 | 0.09 | 24 | 96 | 5.6 | 6.1 | 94 | 96 |
| Trifluralin | y = 30,967x + 967 | 0.9996 | 0.02 | 25 | 96 | 6.9 | 8.0 | 97 | 99 |
| 4-octylphenol | y = 30,367x + 784 | 0.9997 | 0.02 | 24 | 96 | 6.4 | 7.3 | 98 | 100 |
| Tonalide | y = 29,633x + 518 | 0.9998 | 0.01 | 23 | 96 | 7.3 | 8.2 | 92 | 98 |
| Deltamethrin | y = 2321x + 418 | 0.9994 | 0.33 | 30 | 97 | 5.7 | 7.1 | 96 | 98 |

[a] Abbreviations: *LOD*: limit of detection, *EF*: Enrichment factor, *E%*: extraction efficiency, *RSD*: relative standard deviation.

Table 3. Comparison of the performance of the developed procedure with methods reported in literature [a,b].

| Analytes | Matrix | Method | Material | Sample Volume (mL) | Elution Volume (mL) | Equipment | RR (%) | RSD (%) | Ref. |
| --- | --- | --- | --- | --- | --- | --- | --- | --- | --- |
| BP8 | Swimming pool water | SPE | C18 | 100 | 3 | GC-MS | - | <3% (CV) | [27] |
| BPB | River water | SPE | Oasis HLB | 500 | 4 | HPLC-MS/MS | 92.9 | 15.5 | [28] |
| FEN and FLU | Plasma and urine | MSPE | $G/Fe_3O_4$ | 5 | 0.5 | HPLC-PDA | 97.5–102 | 2–4.2 | [29] |
| CUM and 4-OP | Water | SPE | Oasis HLB | 500 | 6 | HPLC-MS/MS | - | - | [30] |
| CLP, DEL and TRIF | Wetland sediments | SPE | Oasis HLB | 500 | 5 | GC-ECD | 69–101 | <15 | [31] |
| FEN, BPB, BP8, CUM, FLU, CLP, TRIF, 4-OP, TON and DEL | Lake water | SPE | MUF | 25 | 5 | HPLC-PDA | 92–100 | 5.6–8.4 | This work |

[a] Methods for the determination of the analytes using sample processing and chromatographic separation are compared. [b] Abbreviations: *RR*, relative recovery; *RSD*, relative standard deviation; *CV*, coefficient of variation; FEN, fenbufen; BPB, butylparaben; BP8, benzophenone-8; CUM, cumylphenol; FLU, flurbiprofen; CLP, chlorpyrifos; TRIF, trifluralin; 4-OP, 4-octylphenol; TON, tonalide; DEL, deltamethrin; SPE, solid-phase extraction; MSPE, magnetic solid-phase extraction; MUF, melamine urea formaldehyde.

## 3. Materials and Methods

### 3.1. Chemical and Reagents

The reagents were at least of analytical grade and were purchased from Sigma-Aldrich (Steinheim, Germany). Stock standard solutions of the analytes, fenbufen (FEN), butylparaben (BPB), benzophenone 8 (BP8), 4-cumylphenol (CUM), flurbiprofen (FLU), chlorpyrifos (CLP), trifluralin (TRIF), 4-octylphenol (4-OP), tonalide (TON), and deltamethrin (DEL) were prepared in methanol at a concentration ranging from 3 to 20 g L$^{-1}$ and were stored at $-18$ °C. Working solutions were prepared daily in double distilled water (DDW). Melamine sponges were bought from a local store in Ioannina (Greece).

### 3.2. Synthesis of Urea-Formaldehyde Sponges

Melamine sponges were firstly cut into cubes of 1 cm$^3$ (1 $\times$ 1 $\times$ 1 cm), rinsed with ethanol and left to dry. The cubes were coated with urea-formaldehyde (UF) oligomers in a simple procedure. In brief, 0.132 g of urea (2.2 mmol) dissolved in 100 mL of DDW was mixed with 66.1 µL of formaldehyde (2.2 mmol) and the pH value was adjusted to 10 with NaOH 1 M. Then, five sponge cubes of ca. 10 mg each were added to the solution and stirred at 85 °C, for 2.5 h. The temperature was decreased to 45 °C and 2.0 mL of methanol was added dropwise to cease the polymerization process and the solution was kept under stirring for a further 10 min. Finally, the solution was cooled to room temperature and the MUF cubes were removed, squeezed to remove excess solution and washed with DDW and ethanol. After the washing step, cubes were dried at 60 °C for 8 h.

### 3.3. Apparatus

The morphology of MUF cubes was observed by scanning electron microscopy (SEM). Images were obtained using a JEOL JSM 6300 microscope at the Central Service for Research Support of the University of Córdoba. Chromatographic analysis was carried out on a Shimadzu (Kyoto, Japan) HPLC system coupled to a SPD-M20A Diode Array Detector (DAD). The column used for the separation was a Hypersil ODS (250 $\times$ 4.6 mm, 5 µm particle size) from Thermo Fisher Scientific (Waltham, MA, USA), kept at 25 °C in a CTO 10AS column oven. Samples were injected using a Rheodyne injector with a sample loop of 20 µL volume. The mobile phase consisted of water (A) and acetonitrile (B), containing 0.1% (v/v) formic acid. The analytes were separated following a gradient elution program from 55% to 80% B in 30 min and then to 85% in 5 min. The total chromatogram time was 35 min. Mobile phase was delivered using a LC20AD pump, at a flow rate of 0.8 mL min$^{-1}$. The detector was set at a wavelength range of 200–360 nm. Data acquisition and processing were carried out using a LC-solution software version 1.21. FTIR spectra were recorded on a Spectrum Two FTIR using an attenuated total reflectance accessory (PerkinElmer, Cambridge, MA, USA). Approximate contact angle measurements were carried out by analyzing various photographs of water droplets on MUF cubes using the Corel Draw X6 software.

### 3.4. SPE Procedure

For the extraction of the analytes, three MUF cubes of the dimensions mentioned above were placed in a SPE cartridge (inner diameter: 0.8 cm), which was attached to a flow control valve (Luer stopcock). The extraction unit accommodated the need for the extraction and final elution step of the analytes and allowed the interaction of sorbent with the sample, at a controlled flow rate of 0.6 mL min$^{-1}$. Before extraction, the cubes were preconditioned with 3 mL of methanol followed by 6 mL of DDW acidified to pH 4.5 with hydrochloric acid. A sample aliquot of 25 mL, the pH of which was previously adjusted to 4.5, was passed through the extraction unit, at a rate of ca. 0.5 mL min$^{-1}$. Then, 5 mL of DDW water (pH 4.5) was passed through the cubes, under the same flow rate as in the cleaning step and the cubes were dried using a nitrogen stream. Finally, the analytes were eluted using

2.0 mL of methanol, which was collected and passed through the sponges three times, evaporated up to 100 µL, under a gentle nitrogen stream and was injected into the HPLC-DAD system.

## 4. Conclusions

In this study, a novel functionalization procedure of MeS is applied. Adopting a base-catalyzed condensation step, which is fundamentally different from that for the typical UF resin synthesis, we obtained UF co-oligomers, that were developed on the surface of MeS. The resulting MUF cubes were highly hydrophobic, owing to the procedure by which the UF oligomers were formed. By altering the parameters that affect the synthesis, the hydrophilic/lipophilic balance of the functionalized cubes can be tuned. Furthermore, capitalizing on the hydrophobic properties of the synthesized MUF cubes we developed an SPE procedure, suitable for hydrophobic analytes. The analytical method achieves low LOQs, satisfactory recoveries, good reproducibility and wide linear ranges. Overall, the developed material can serve as an excellent alternative sorbent for classical SPE procedures.

**Supplementary Materials:** The following are available online. Table S1: Total amounts of urea and formaldehyde used for the synthesis of MUF cubes and the respective total extraction yields of the examined analytes, Figure S1: SEM micrographs of (A) MeS and (B) MUF cubes, Figure S2: Effect of various sample volumes (spiked with constant amount or the same concentration of analytes) on the extraction efficiency of the method, Figure S3: Chromatograms (at 285 nm) of a blank lake water sample (lower chromatogram) and a lake water sample spiked with 50 µg L$^{-1}$ of the analytes (upper chromatogram). Abbreviations: FEN, fenbufen; BPB, butylparaben; BP8, benzophenone-8; CUM, cumylphenol; FLU, flurbiprofen; CLP, chlorpyrifos; TRIF, trifluralin; 4-OP, 4-octylphenol; TON, tonalide; DEL, deltamethrin.

**Author Contributions:** Conceptualization, C.S.; Formal analysis, R.L. and S.C.; Investigation, T.C. and M.T.G.-V.; Methodology, T.C. and M.T.G.-V.; Supervision, C.S.; Writing—original draft, C.S.; Writing—review & editing, C.S., R.L. and S.C.

**Funding:** M.T.G.-V. was funded by the Spanish Ministry of Education (ref BES-2015-071421).

**Acknowledgments:** We thank the Central Service for Research Support from the University of Córdoba. Images were obtained using a JEOL JSM 6300 microscope in the Central Service for Research Support (SCAI) of the University of Córdoba.

**Conflicts of Interest:** The authors declare no conflict of interest.

## References

1. Chatzimitakos, T.; Stalikas, C. Carbon-Based Nanomaterials Functionalized with Ionic Liquids for Microextraction in Sample Preparation. *Separations* **2017**, *4*, 14. [CrossRef]

2. Buszewski, B.; Szultka, M. Past, Present, and Future of Solid Phase Extraction: A Review. *Crit. Rev. Anal. Chem.* **2012**, *42*, 198–213. [CrossRef]

3. Andrade-Eiroa, A.; Canle, M.; Leroy-Cancellieri, V.; Cerdà, V. Solid-phase extraction of organic compounds: A critical review (Part I). *Trends Anal. Chem.* **2016**, *80*, 641–654. [CrossRef]

4. Płotka-Wasylka, J.; Szczepańska, N.; de la Guardia, M.; Namieśnik, J. Modern trends in solid phase extraction: New sorbent media. *Trends Anal. Chem.* **2016**, *77*, 23–43. [CrossRef]

5. Pham, V.H.; Dickerson, J.H. Superhydrophobic Silanized Melamine Sponges as High Efficiency Oil Absorbent Materials. *ACS Appl. Mater. Interfaces* **2014**, *6*, 14181–14188. [CrossRef] [PubMed]

6. Reyssat, M.; Richard, D.; Clanet, C.; Quéré, D. Dynamical superhydrophobicity. *Faraday Discuss.* **2010**, *146*, 19–33. [CrossRef] [PubMed]

7. Vinhal, J.O.; Lima, C.F.; Cassella, R.J. Polyurethane foam loaded with sodium dodecylsulfate for the extraction of 'quat' pesticides from aqueous medium: Optimization of loading conditions. *Ecotoxicol. Environ. Saf.* **2016**, *131*, 72–78. [CrossRef] [PubMed]

8. Moawed, E.A.; Abulkibash, A.B.; El-Shahat, M.F. Synthesis of tannic acid azo polyurethane sorbent and its application for extraction and determination of atrazine and prometryn pesticides in foods and water samples. *Environ. Nanotechnol. Monit. Manag.* **2015**, *3*, 61–66. [CrossRef]

9. Moawed, E.A.; Radwan, A.M. Application of acid modified polyurethane foam surface for detection and removing of organochlorine pesticides from wastewater. *J. Chromatogr. B* **2017**, *1044–1045*, 95–102. [CrossRef] [PubMed]

10. Yang, Y.; Deng, Y.; Tong, Z.; Wang, C. Multifunctional foams derived from poly(melamine formaldehyde) as recyclable oil absorbents. *J. Mater. Chem. A* **2014**, *2*, 9994–9999. [CrossRef]
11. Hou, K.; Jin, Y.; Chen, J.; Wen, X.; Xu, S.; Cheng, J.; Pi, P. Fabrication of superhydrophobic melamine sponges by thiol-ene click chemistry for oil removal. *Mater. Lett.* **2017**, *202*, 99–102. [CrossRef]
12. Lei, Z.; Zhang, G.; Ouyang, Y.; Liang, Y.; Deng, Y.; Wang, C. Simple fabrication of multi-functional melamine sponges. *Mater. Lett.* **2017**, *190*, 119–122. [CrossRef]
13. Huang, J.; Xu, Y.; Zhang, X.; Lei, Z.; Chen, C.; Deng, Y.; Wang, C. Polyethylenimine and dithiocarbamate decorated melamine sponges for fast copper(II) ions removal from aqueous solution. *Appl. Surf. Sci.* **2018**, *445*, 471–477. [CrossRef]
14. Deng, C.-H.; Gong, J.-L.; Zhang, P.; Zeng, G.-M.; Song, B.; Liu, H.-Y. Preparation of melamine sponge decorated with silver nanoparticles-modified graphene for water disinfection. *J. Colloid Interface Sci.* **2017**, *488*, 26–38. [CrossRef] [PubMed]
15. Peng, M.; Chen, G.; Zeng, G.; Chen, A.; He, K.; Huang, Z.; Hu, L.; Shi, J.; Li, H.; Yuan, L.; et al. Superhydrophobic kaolinite modified graphene oxide-melamine sponge with excellent properties for oil-water separation. *Appl. Clay Sci.* **2018**, *163*, 63–71. [CrossRef]
16. Zhao, J.; Guo, Q.; Wang, X.; Xie, H.; Chen, Y. Recycle and reusable melamine sponge coated by graphene for highly efficient oil-absorption. *Colloids Surf. A* **2016**, *488*, 93–99. [CrossRef]
17. Chen, J.; You, H.; Xu, L.; Li, T.; Jiang, X.; Li, C.M. Facile synthesis of a two-tier hierarchical structured superhydrophobic-superoleophilic melamine sponge for rapid and efficient oil/water separation. *J. Colloid Interface Sci.* **2017**, *506*, 659–668. [CrossRef] [PubMed]
18. Chatzimitakos, T.; Samanidou, V.; Stalikas, C.D. Graphene-functionalized melamine sponges for microextraction of sulfonamides from food and environmental samples. *J. Chromatogr. A* **2017**, *1522*, 1–8. [CrossRef] [PubMed]
19. Chatzimitakos, T.G.; Stalikas, C.D. Melamine sponge decorated with copper sheets as a material with outstanding properties for microextraction of sulfonamides prior to their determination by high-performance liquid chromatography. *J. Chromatogr. A* **2018**, *1554*, 28–36. [CrossRef] [PubMed]
20. Philbrook, A.; Blake, C.J.; Dunlop, N.; Easton, C.J.; Keniry, M.A.; Simpson, J.S. Demonstration of co-polymerization in melamine–urea–formaldehyde reactions using 15N NMR correlation spectroscopy. *Polymer* **2005**, *46*, 2153–2156. [CrossRef]
21. Cao, J.; Yan, H.; Shen, S.; Bai, L.; Liu, H.; Qiao, F. Hydrophilic molecularly imprinted melamine-urea-formaldehyde monolithic resin prepared in water for selective recognition of plant growth regulators. *Anal. Chim. Acta* **2016**, *943*, 136–145. [CrossRef] [PubMed]
22. Dunky, M. Urea–formaldehyde (UF) adhesive resins for wood. *Int. J. Adhes. Adhes.* **1998**, *18*, 95–107. [CrossRef]
23. Li, T.; Cao, M.; Liang, J.; Xie, X.; Du, G. New Mechanism Proposed for the Base-Catalyzed Urea–Formaldehyde Condensation Reactions: A Theoretical Study. *Polymers* **2017**, *9*, 203. [CrossRef]
24. Chatzimitakos, T.G.; Pierson, S.A.; Anderson, J.L.; Stalikas, C.D. Enhanced magnetic ionic liquid-based dispersive liquid-liquid microextraction of triazines and sulfonamides through a one-pot, pH-modulated approach. *J. Chromatogr. A* **2018**, *1571*, 47–54. [CrossRef] [PubMed]
25. Wells, M. Handling large volume samples: Applications of SPE to environmental matrices. In *Solid-Phase Extraction: Principles, Techniques and Applications*; CRC Press: Boca Raton, FL, USA, 2000; pp. 97–119, ISBN 9780824700218.
26. Chatzimitakos, T.; Binellas, C.; Maidatsi, K.; Stalikas, C. Magnetic ionic liquid in stirring-assisted drop-breakup microextraction: Proof-of-concept extraction of phenolic endocrine disrupters and acidic pharmaceuticals. *Anal. Chim. Acta* **2016**, *910*, 53–59. [CrossRef] [PubMed]
27. Lempart, A.M.; Kudlek, E.A.; Lempart, M.; Dudziak, M. The Presence of Compounds from the Personal Care Products Group in Swimming Pool Water. *J. Ecol. Eng.* **2018**, *19*, 29–37. [CrossRef]
28. González-Mariño, I.; Quintana, J.B.; Rodríguez, I.; Cela, R. Simultaneous determination of parabens, triclosan and triclocarban in water by liquid chromatography/electrospray ionisation tandem mass spectrometry. *Rapid Commun. Mass Spectrom.* **2009**, *23*, 1756–1766. [CrossRef] [PubMed]

29. Ferrone, V.; Carlucci, M.; Ettorre, V.; Cotellese, R.; Palumbo, P.; Fontana, A.; Siani, G.; Carlucci, G. Dispersive magnetic solid phase extraction exploiting magnetic graphene nanocomposite coupled with UHPLC-PDA for simultaneous determination of NSAIDs in human plasma and urine. *J. Pharm. Biomed. Anal.* **2018**, *161*, 280–288. [CrossRef] [PubMed]

30. Benijts, T.; Dams, R.; Lambert, W.; De Leenheer, A. Countering matrix effects in environmental liquid chromatography–electrospray ionization tandem mass spectrometry water analysis for endocrine disrupting chemicals. *J. Chromatogr. A* **2004**, *1029*, 153–159. [CrossRef] [PubMed]

31. Xue, N.; Li, F.; Hou, H.; Li, B. Occurrence of endocrine-disrupting pesticide residues in wetland sediments from Beijing, China. *Environ. Toxicol. Chem.* **2009**, *27*, 1055–1062. [CrossRef] [PubMed]

**Sample Availability:** Samples of the compounds are available from the authors..

*molecules*

MDPI

*Article*

# Determination of Enantiomeric Excess by Solid-Phase Extraction Using a Chiral Metal-Organic Framework as Sorbent

Jun-Hui Zhang [1], Bo Tang [1], Sheng-Ming Xie [1], Bang-Jin Wang [1], Mei Zhang [2], Xing-Lian Chen [3], Min Zi [1] and Li-Ming Yuan [1,*]

[1]   Department of Chemistry, Yunnan Normal University, Kunming 650500, China; zjh19861202@126.com (J.-H.Z.); tangbobengbu@163.com (B.T.); xieshengming_2006@163.com (S.-M.X.); wangbangjin711@163.com (B.-J.W.); zimin48@126.com (M.Z.)
[2]   College of Pharmaceutical Science, Yunnan University of Traditional Chinese Medicine, Kunming 650500, China; meizhang213@163.com
[3]   Institute of Quality Standards and Testing Technology, Yunnan Academy of Agricultural Sciences, Kunming 650224, China; chenxinglian@126.com
*    Correspondence: yuan_limingpd@126.com or yuan_liming@hotmail.com; Tel./Fax: +86-871-6594-1088

Received: 17 August 2018; Accepted: 22 October 2018; Published: 29 October 2018

**Abstract:** Metal-organic frameworks (MOFs) have recently attracted considerable attention because of their fascinating structures and intriguing potential applications in diverse areas. In this study, we developed a novel method for determination of enantiomeric excess (*ee*) of (±)-1,1'-bi-2-naphthol by solid-phase extraction (SPE) using a chiral MOF, [Co(L-tyr)]$_n$(L-tyrCo), as sorbent. After optimization of the experimental conditions, a good linear relationship between the *ee* and the absorbance of the eluate ($R^2$ = 0.9984) was obtained and the standard curve was established at the concentration of 3 mmol L$^{-1}$. The *ee* values of (±)-1,1'-bi-2-naphthol samples can be rapidly calculated using the standard curve after determination of the absorbance of the eluate. The method showed good accuracy, with an average error of 2.26%, and is promising for *ee* analysis.

**Keywords:** metal-organic frameworks; enantiomeric excess; chiral compounds; solid-phase extraction

---

## 1. Introduction

Optical purity is important in many industries, particularly in the development and production of pharmaceuticals. In general, one enantiomer of chiral drugs is safe and effective, while the other is may be ineffective or even toxic [1]. Consequently, the determination of enantiomeric excess (*ee*) of chiral compounds is very important. At present, chiral chromatographic methods, such as high performance liquid chromatography (HPLC) and gas chromatography (GC) using chiral stationary phases (CSPs), are most commonly used to determine *ee* values. However, these methods have some drawbacks, such as the high cost of chiral columns and long analysis times. Therefore, the development of simple and fast methods for *ee* analysis has great value. A number of techniques for rapid *ee* analysis have been developed, including circular dichroism [2,3], UV and colorimetric methods [4,5], fluorescence spectroscopy [6,7], mass spectrometry [8], and molecularly imprinted polymer based assays [9].

Solid-phase extraction (SPE) is a valuable sample pretreatment technique that enables concentration and purification of analytes from complex matrices. It has been widely used for its simplicity, rapidity, low cost, and compatibility with different detection techniques in both on-line and off-line modes [10,11]. The adsorbent is the core in SPE. High enrichment efficiency and selectivity of SPE are particularly dependent on the adsorbent. Thus, development of novel and efficient adsorbents for SPE is an ongoing research interest.

Metal-organic frameworks (MOFs) are an emerging class of crystalline porous materials that combine metal nodes with organic linkers through coordination bonds [12,13]. MOFs have attracted considerable attention in recent years because of their fascinating structures and outstanding properties, such as permanent nanoscale porosity, high surface area, good chemical stability, rich topology, and tunable pore size. These unique features make MOFs attractive as advanced media for separation [14,15], gas adsorption and storage [16,17], catalysis [18,19], sensing [20,21], imaging [22], and other applications. To date, a number of MOFs, such as MIL-53 [23,24], MIL-100 [23], MIL-101 [23,25,26], MOF-199 [27,28], ZIF-8 [29,30], ZIF-67 [31], ZIF-90 [32], MIL-88B [33], MAF-X8 [34], MOF-5 [35], UIO-66 [36], and UIO-67 [37], have been used as adsorbents for SPE or solid-phase microextraction (SPME) and shown much promise [14,38,39].

The chiral 3D MOF, [Co(L-tyr)]$_n$(L-tyrCo), composed of metal dimers (Co$^{II}$) linked via L-tyrosine ligand (Figure 1a) [40]. In this paper, we report the determination of *ee* values of (±)-1,1′-bi-2-naphthol (Figure 1b) by SPE using [Co(L-tyr)]$_n$(L-tyrCo) as the adsorbent. The experimental results showed that the *ee* values of (±)-1,1′-bi-2-naphthol were linearly correlated with the absorbance of the eluates, and a standard curve could be established for *ee* analysis. The *ee* values of (±)-1,1′-bi-2-naphthol samples can be rapidly measured using the standard curve by determining the absorbance of the eluates after SPE. The method has advantages that include of good accuracy, low cost, simple operation, and suitability for batch analysis.

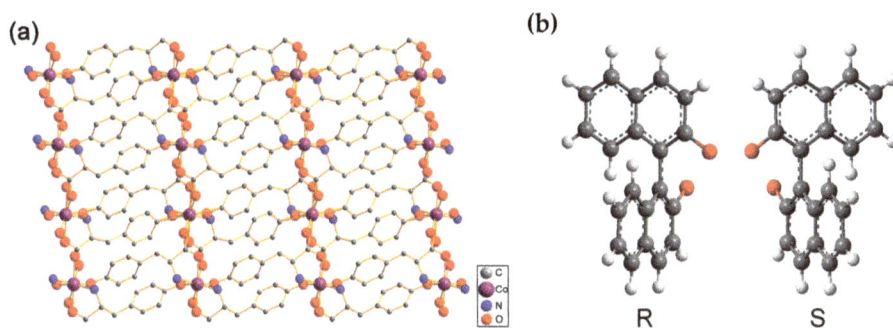

**Figure 1. (a)** 2D crystal structure of the chiral metal-organic framework (MOF), [Co(L-tyr)]$_n$(L-tyrCo); **(b)** the structures of 1,1′-bi-2-naphthol enantiomers.

## 2. Results and Discussion

### 2.1. Characterization of the Synthesized [Co(L-tyr)]$_n$(L-tyrCo)

The synthesized [Co(L-tyr)]$_n$(L-tyrCo) was characterized by powder X-ray diffraction (PXRD). As shown in Figure 2a, the experimental PXRD pattern of the synthesized [Co(L-tyr)]$_n$(L-tyrCo) crystals was in good agreement with that simulated from single crystal data, indicating successful synthesis. After several extraction experiments and soaking in a large volume of methanol and *n*-hexane/isopropanol (70/30, *v*/*v*) for 48 h, the PXRD pattern was not significantly changed, suggesting good stability of the MOF, and that it would not be dissolved during the experiments Figure 2a, (3) and (4). The scanning electron microscopy (SEM) image showed the selected crystals of [Co(L-tyr)]$_n$(L-tyrCo) with an average size of about 10 μm (Figure 2b).

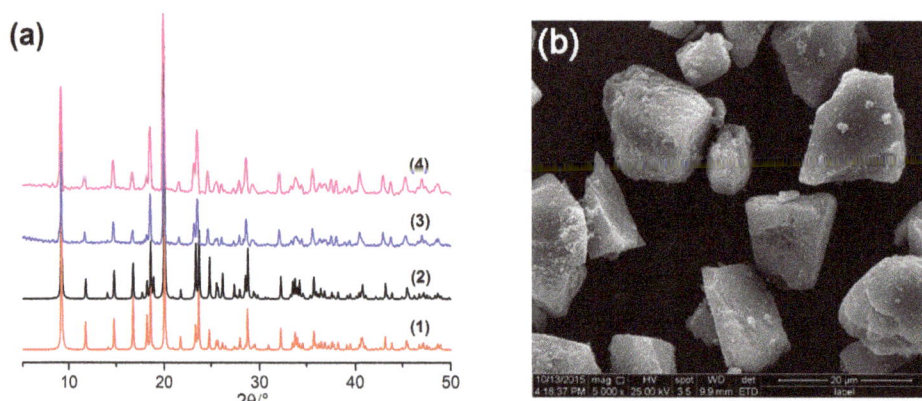

**Figure 2.** (**a**) Powder X-ray diffraction (PXRD) patterns of [Co(L-tyr)]$_n$(L-tyrCo): (1) simulated from single crystal X-ray diffraction data, (2) as-synthesized, (3) after the experiment, (4) after soaking in a large volume of methanol and n-hexane/isopropanol (70/30, $v/v$) for 48 h; (**b**) SEM image of the selected [Co(L-tyr)]$_n$(L-tyrCo) particles.

## 2.2. Selection of UV Wavelength for Analysis of Eluate

To obtain the highest sensitivity, the optimum UV absorption wavelength of the analyte should be determined. The UV-visible absorption spectrum of (±)-1,1′-bi-2-naphthol in methanol was measured over 200–400 nm. As can be seen in Figure 3, the solution of (±)-1,1′-bi-2-naphthol showed strong absorptions at 278 and 335 nm. In principle, both of these wavelengths could be used to determine the absorbance of the eluate. However, some impurities in the eluent absorbed at around 271 nm, which would affect the accuracy of the experimental result. In contrast, impurities in the eluent had relatively low absorption at 335 nm, so this wavelength was used in subsequent experiments.

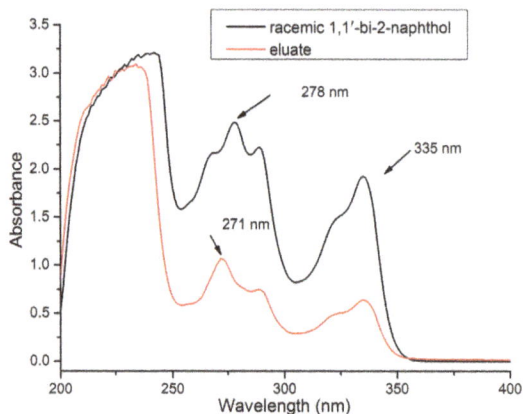

**Figure 3.** UV-visible spectra of racemic (±)-1,1′-bi-2-naphthol and eluate.

## 2.3. Optimization of SPE Conditions

The sorption of (±)-1,1′-bi-2-naphthol on a SPE cartridge packed with [Co(L-tyr)]$_n$(L-tyrCo) can be influenced by many experimental conditions, such as solvent composition, sample solution volume, analyte concentration, and the type and volume of eluent. To obtain good extraction efficiency, these parameters were optimized.

### 2.3.1. Effect of Sample Solvent Composition

The sample solvent composition will influence adsorption of the analyte on SPE adsorbent. In this study, (±)-1,1'-bi-2-naphthol was dissolved in mixtures of *n*-hexane/isopropanol mixtures at different ratios (90/10, 80/20, 70/30, 60/40, 50/50, 40/60, 30/70, 20/80, 10/90, and 0/100; *v/v*). As can be seen in Figure 4, the maximum absorbance of the eluate was obtained when using *n*-hexane/ isopropanol = 70/30 (*v/v*) as solvent. This solvent mixture was therefore chosen for the following experiments.

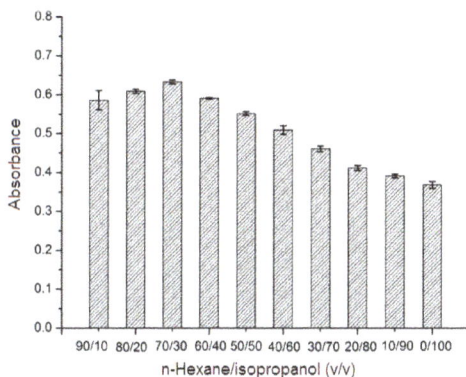

**Figure 4.** Effect of solvent composition. Analyte concentration: 3.0 mmol $L^{-1}$; sample volume: 2.0 mL; and elution solvent: methanol (4 mL).

### 2.3.2. Effect of Sample Solution Volume

The adsorption efficiency can also be influenced by the volume of the sample solution. Volumes of sample solution (3 mmol $L^{-1}$) in the range of 1.0–3.0 mL were passed through the SPE cartridge, and the adsorbed analytes were eluted with methanol (4 mL). As shown in Figure 5, the absorbance of the eluate increased as the sample solution volume was increased from 1.0 mL to 2.0 mL, but remained stable as the volume was increased to 3.0 mL. Thus, the optimum volume was 2.0 mL.

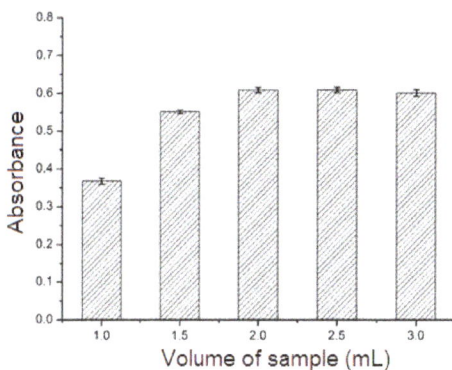

**Figure 5.** Effect of sample solution volume. Analyte concentration: 3.0 mmol $L^{-1}$; elution solvent: methanol (4 mL).

### 2.3.3. Effect of Analyte Concentration

Five different concentrations (1.0, 2.0, 3.0, 4.0, and 5.0 mmol $L^{-1}$) of (±)-1,1'-bi-2-naphthol were investigated. As shown in Figure 6, the absorbance of the eluate increased as the analyte concentration was increased from 1.0 to 3.0 mmol $L^{-1}$, with no further increase as the concentration was raised to 5.0 mmol $L^{-1}$. The optimal concentration was therefore 3.0 mmol $L^{-1}$.

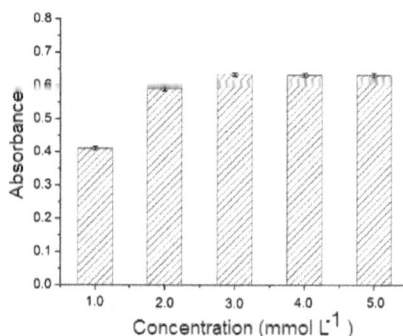

**Figure 6.** Effect of analyte concentration. Sample volume: 2.0 mL; and elution solvent: methanol (10 mL).

### 2.3.4. Effect of Type and Volume of Elution Solvent

The nature of the elution solvent plays an important role in the effective desorption of analyte from adsorbent. The influence of elution solvent on desorption was evaluated under the optimum conditions using six common organic solvents: Methanol, Ethanol, Isopropanol, Acetonitrile, Ethyl acetate, and *n*-Hexane. As shown in Figure 7a, the maximum absorbance of the eluate was obtained when using methanol as the desorption solvent. Therefore, methanol was adopted as the optimal desorption solvent. The volume of methanol was also optimized because the sorption analyte cannot be completely eluted by a small amount of methanol, and long volatilization time will be taken when excessive methanol is used. As presented in Figure 7b, the absorbance of the eluate increased as the volume of methanol was increased from 1 to 4 mL, and no significant changes were observed as the volume was increased further. Consequently, 4 mL of methanol was sufficient for the analytes to be completely eluted and was chosen as the optimized volume.

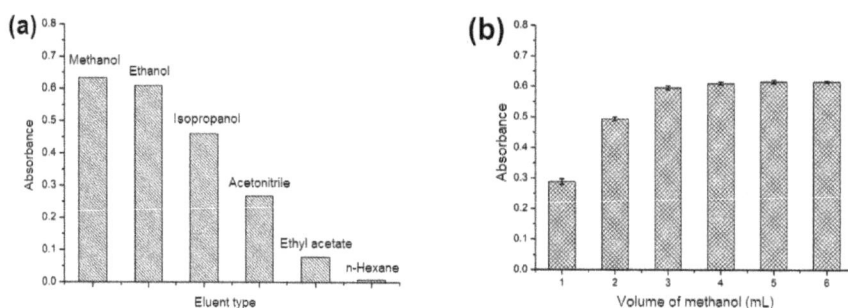

**Figure 7.** Optimization of (**a**) elution solvent; (**b**) elution volume. Analyte concentration: 3.0 mmol $L^{-1}$; sample volume: 2.0 mL.

### 2.4. Reproducibility and Reusability of the SPE Column

To evaluate the reproducibility of the SPE column, the sample solution (3 mmol $L^{-1}$, 2 mL) was added to each of 30 SPE cartridges filled with different batches of $[Co(L\text{-tyr})]_n$(L-tyrCo) particles. As shown in Figure 8a, the differences in the absorbance values of the eluates among the 30 columns were very small, indicating the good reproducibility of the SPE column. The reusability of the SPE column was also investigated. After each SPE experiment, the column was washed sequentially with methanol (10 mL) and *n*-hexane/isopropanol (70/30, *v/v*, 10 mL), and then reused for the next extraction cycle. As presented in Figure 8b, no obvious change in absorbance of the eluate was observed after five SPE cycles, indicating good reusability of the SPE column.

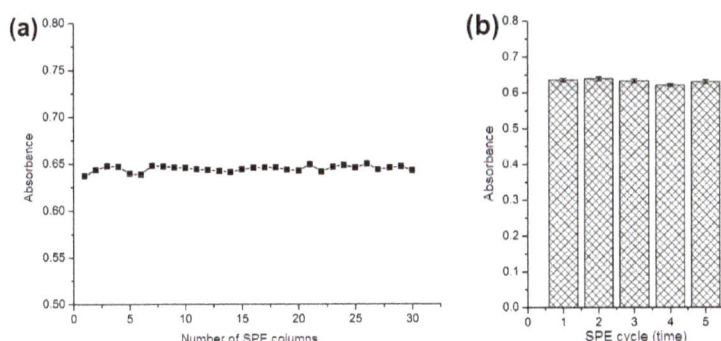

**Figure 8.** (**a**) Reproducibility and (**b**) reusability of the SPE column.

*2.5. Method Validation*

Solutions of (±)-1,1′-bi-2-naphthol with varying *ee* values (from −100% to +100%) in *n*-hexane/isopropanol (70:30, *v*/*v*, 3 mmol $L^{-1}$) were prepared by mixing solutions of (*R*)-1,1′-bi-2-naphthol and (*S*)-1,1′-bi-2-naphthol. The solutions (2.0 mL) were passed through the SPE columns under the optimum conditions. The absorbance values of the eluates were determined by UV-visible spectrophotometry (Figure 9a), and the absorbance data at 335 nm were collected. All of the data were fitted by computer and all possible functions tried. An excellent linear relationship between *ee* and the absorbance of the eluate ($R^2$ = 0.9984) was obtained at the concentration of 3 mmol $L^{-1}$ (Figure 9b), which enabled the calculation of *ee* values for different (±)-1,1′-bi-2-naphthol samples. The selectivity of the MOF for each enantiomer was studied. The adsorption of (*S*)-1,1′-bi-2-naphthol by the MOF was greater than that of (*R*)-1,1′-bi-2-naphthol. The absolute recoveries were 62% for (*R*)-1,1′-bi-2-naphthol and 98% for (*S*)-1,1′-bi-2-naphthol under the developed conditions.

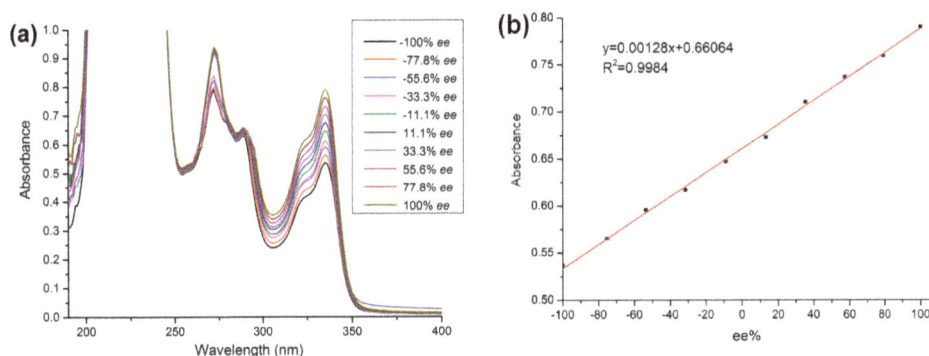

**Figure 9.** (**a**) UV-visible spectra of the eluates after extraction of (±)-1,1′-bi-2-naphthol at various *ee* values; (**b**) standard curve showing the linearity of *ee*% with the absorbance of the eluates.

Given that an enantiomerically pure of L-tyrosine was used as ligand in the synthesis, the synthesized [Co(L-tyr)]$_n$(L-tyrCo) is chiral. The 2D structure of [Co(L-tyr)]$_n$(L-tyrCo) showing cobalt oxide layers pillared with L-tyrosine linkers and produced large chiral pores (Figure 1a). The influence of the chiral microenvironment on the enantioselective adsorption is complicated. It is difficult to completely understand the mechanism of interaction between the MOF and the two enantiomers. The reason for the selective adsorption of (*S*)-1,1′-bi-2-naphthol may be the high degree of space matching between (*S*)-1,1′-bi-2-naphthol and the chiral structures of the crystal resulting in different forces.

To verify the accuracy of the methodology, solutions of (±)-1,1′-bi-2-naphthol at a variety of *ee* values were determined by this method. The absorbances of the eluates were determined and the *ee* values were calculated using the standard curve (Table 1). Compared with the actual *ee* values measured by chiral HPLC, this method had an average error of 2.26%. This method can therefore be used for evaluation of the *ee* of (±)-1,1′-bi-2-naphthol with good accuracy.

**Table 1.** Enantiomeric excess values of (±)-1,1′-bi-2-naphthol calculated from the SPE-based assay.

| Sample | Absorbance | Experimental *ee* (%) | Actual *ee* (%) | Absolute Error (%) | Average Error (%) |
|--------|-----------|----------------------|-----------------|--------------------|--------------------|
| 1 | 0.5566 | −81.25 | −84.41 | 3.16 | |
| 2 | 0.6022 | −45.67 | −43.34 | 2.33 | |
| 3 | 0.6582 | −1.93 | −0.87 | 1.06 | 2.26 |
| 4 | 0.7098 | 38.41 | 41.79 | 3.38 | |
| 5 | 0.7621 | 79.24 | 77.89 | 1.35 | |

## 3. Materials and Methods

### 3.1. Reagents and Materials

All chemicals and reagents used in this study were at least of analytical grade. (*R*)-1,1′-Bi-2-naphthol and (*S*)-1,1′-bi-2-naphthol were purchased from Sigma-Aldrich (St. Louis, MO, USA). $Co(CH_3COO)_2 \cdot 4H_2O$, L-tyrosine, and NaOH were purchased from Adamas-beta (Shanghai, China), and were used for the synthesis of MOF, $[Co(L-tyr)]_n$ (L-tyrCo). HPLC grade methanol, ethanol, isopropanol, acetonitrile, ethyl acetate, and *n*-hexane were obtained from TEDIA (Fairfield, OH, USA). Ultrapure water (18.2 Ω cm) was produced with an ELGA LabWater water purification system (High Wycombe, UK). The empty SPE cartridges (0.2000 g, 3 mL, polypropylene) with frits (20 μm porosity) were purchased from SEPAX Technologies Inc. (Suzhou, China).

### 3.2. Instrumentations

Powder X-ray diffraction (PXRD) patterns were recorded on a D/max-3B diffractometer (Tokyo, Japan) using Cu $K_\alpha$ radiation. Scanning electron microscopy (SEM) images were recorded on a FEI Quanta FEG 650 scanning electron microscope (Hillsboro, OR, USA). The absorbance of eluate was determined on a TU-1901 UV/Vis spectrophotometer (Beijing Purkinje General Instrument Co., Ltd., Beijing, China). Commercial ChiralPak OD-H (250 mm × 4.6 mm, Daicel, Shanghai, China) and a Shimadzu HPLC system (Tokyo, Japan) consisting of an LC-15C HPLC pump and SPD-15C UV/Vis detector were used for the actual *ee* analyses.

### 3.3. Synthesis of $[Co(L-tyr)]_n$ (L-tyrCo)

$[Co(L-tyr)]_n$ (L-tyrCo) was synthesized according to the method of Rocha et al. [40]. Typically, $Co(CH_3COO)_2 \cdot 4H_2O$ and L-tyrosine in a molar ratio of 1:2 were mixed with ultrapure water (20 mL) in a Teflon-lined bomb and stirred with a magnetic bar. The pH of the mixture was adjusted to 9~10 by adding aqueous solution of NaOH (1 mol $L^{-1}$). The bomb was sealed and heated at 130 °C for 3 days. After cooling to room temperature, the mixture was filtered and the purple crystals were washed thoroughly with distilled water and ethanol. Finally, the solid was dried at 100 °C.

### 3.4. Preparation of SPE Columns

The synthesized crystals of $[Co(L-tyr)]_n$ (L-tyrCo) were large and heterogeneous, which made them unsuitable for direct use as adsorbent in the preparation of SPE columns. The crystals were therefore milled and a suitable particle size was selected by preparing an ethanol suspension. Subsequently, each SPE cartridge was packed with the selected crystal particles (0.2000 g) using polypropylene frits at each end of the cartridge to keep the packing in place. The outlet tip of the cartridge was connected to a vacuum pump, and methanol was continuously added at the inlet end.

*3.5. SPE Procedures*

Before the extraction experiment, the crystals of the [Co(L-tyr)]$_n$(L-tyrCo) packed cartridge were preconditioned by washing with methanol (5 mL) and activated with sample solvent (5 mL). A known volume of the sample was passed through the activated cartridge by gravity action. To remove residual sample on the surface of the crystals, additional sample solvent (5 mL) was added after the sample solution had completely passed through. Subsequently, the cartridge was eluted with methanol (4 mL). The eluate was volatilized, diluted with methanol to a final volume of 4 Ml, and the absorbance was determined by UV-visible spectrophotometry.

*3.6. Calculation of ee Value*

The *ee* value of the sample was calculated using the following formula:

$$ee\% = \frac{S-R}{S+R} \times 100\%$$

(*R*, *S*: Concentration/peak area, respectively).

Logically, *ee* values of +100% and −100% represent enantiomerically pure of (*S*)-1,1′-bi-2-naphthol and (*R*)-1,1′-bi-2-naphthol, respectively, and an *ee* value of 0% indicates racemate.

## 4. Conclusions

In this work, we reported a novel and simple method for the determination of *ee* value by SPE using a chiral MOF as sorbent. After adsorption and desorption, a standard curve was established that demonstrated a good linear relationship between *ee* and absorbance of the eluate. The *ee* value of (±)-1,1′-bi-2-naphthol can be rapidly calculated after SPE using the standard curve.

**Author Contributions:** J.-H.Z. performed the experiments and wrote the manuscript; B.T. and S.-M.X. synthesized the chiral MOF; B.-J.W. characterized the MOF; M.Z. (Min Zi) analyzed the data; M.Z. (Mei Zhang) and X.-L.C. analyzed the data and helped revise the manuscript; L.-M.Y. designed and supervised the research work.

**Funding:** This research was supported by the National Natural Science Foundation of China (No. 21565032, No. 21705142, No. 21675141 and No. 21665030).

**Conflicts of Interest:** The authors declare no conflict of interest.

## References

1. Maier, N.; Franco, P.; Lindner, W. Separation of enantiomers: Needs, challenges, perspectives. *J. Chromatogr. A* **2001**, *906*, 3–33. [CrossRef]
2. Jo, H.H.; Lin, C.Y.; Anslyn, E.V. Rapid optical methods for enantiomeric excess analysis: From enantioselective indicator displacement assays to exciton-coupled circular dichroism. *Acc. Chem. Res.* **2014**, *47*, 2212–2221. [CrossRef] [PubMed]
3. Dragna, J.M.; Gade, A.M.; Tran, L.; Lynch, V.M.; Anslyn, E.V. Chiral amine enantiomeric excess determination using self-assembled octahedral Fe(II)-imine complexes. *Chirality* **2015**, *27*, 294–298. [CrossRef] [PubMed]
4. Folmer-Andersen, J.F.; Lynch, V.M.; Anslyn, E.V. Colorimetric enantiodiscrimination of α-amino acids in protic media. *J. Am. Chem. Soc.* **2005**, *127*, 7986–7987. [CrossRef] [PubMed]
5. Mei, X.; Wolf, C. Determination of enantiomeric excess and concentration of unprotected amino acids, amines, amino alcohols, and carboxylic acids by competitive binding assays with a chiral scandium complex. *J. Am. Chem. Soc.* **2006**, *128*, 13326–13327. [CrossRef] [PubMed]
6. Korbel, G.A.; Lalic, G.; Shair, M.D. Reaction microarrays: A method for rapidly determining the enantiomeric excess of thousands of samples. *J. Am. Chem. Soc.* **2001**, *123*, 361–362. [CrossRef] [PubMed]
7. Li, P. Fluorescence of organic molecules in chiral recognition. *Chem. Rev.* **2004**, *104*, 1687–1716.

8.  Guo, J.; Wu, J.; Siuzdak, G.; Finn, M.G. Measurement of enantiomeric excess by kinetic resolution and mass spectrometry. *Angew. Chem. Int. Ed.* **1999**, *38*, 1755–1758. [CrossRef]
9.  Chen, Y.Z.; Shimizu, K.D. Measurement of enantiomeric excess using molecularly imprinted polymers. *Org. Lett.* **2002**, *4*, 2937–2940. [CrossRef] [PubMed]
10. Li, Y.; Lu, P.; Cheng, J.; Zhu, X.; Guo, W.; Liu, L.; Wang, Q.; He, C.; Liu, S. Novel microporous β-cyclodextrin polymer as sorbent for solid-phase extraction of bisphenols in water samples and orange juice. *Talanta* **2018**, *187*, 207–215. [CrossRef] [PubMed]
11. Płotka-Wasylka, J.; Szczepańska, N.; de la Guardia, M.; Namieśnik, J. Modern trends in solid phase extraction: New sorbent media. *TrAC Trends Anal. Chem.* **2016**, *77*, 23–43. [CrossRef]
12. Yaghi, O.M.; O'Keeffe, M.; Ockwig, N.W.; Chae, H.K.; Eddaoudi, M.; Kim, J. Reticular synthesis and the design of new materials. *Nature* **2003**, *423*, 705–714. [CrossRef] [PubMed]
13. Férey, G. Hybrid porous solids: Past, present, future. *Chem. Soc. Rev.* **2008**, *37*, 191–214. [CrossRef] [PubMed]
14. Gu, Z.Y.; Yang, C.X.; Chang, N.; Yan, X.P. Metal-organic frameworks for analytical chemistry: From sample collection to chromatographic separation. *Acc. Chem. Res.* **2012**, *45*, 734–745. [CrossRef] [PubMed]
15. Li, J.R.; Sculley, J.; Zhou, H.C. Metal-organic frameworks for separations. *Chem. Rev.* **2012**, *112*, 869–932. [CrossRef] [PubMed]
16. Murray, L.J.; Dinca, M.; Long, J.R. Hydrogen storage in metal-organic frameworks. *Chem. Soc. Rev.* **2009**, *38*, 1294–1314. [CrossRef] [PubMed]
17. Li, J.R.; Kuppler, R.J.; Zhou, H.C. Selective gas adsorption and separation in metal-organic frameworks. *Chem. Soc. Rev.* **2009**, *38*, 1477–1504. [CrossRef] [PubMed]
18. Liu, Y.; Xuan, W.M.; Cui, Y. Engineering homochiral metal-organic frameworks for heterogeneous asymmetric catalysis and enantioselective separation. *Adv. Mater.* **2010**, *22*, 4112–4135. [CrossRef] [PubMed]
19. Lee, J.; Farha, O.K.; Roberts, J.; Scheidt, K.A.; Nguyen, S.T.; Hupp, J.T. Metal-organic framework materials as catalysts. *Chem. Soc. Rev.* **2009**, *38*, 1450–1459. [CrossRef] [PubMed]
20. Allendorf, M.D.; Bauer, C.A.; Bhakta, R.K.; Houk, R.J.T. Luminescent metal-organic frameworks. *Chem. Soc. Rev.* **2009**, *38*, 1330–1352. [CrossRef] [PubMed]
21. Cui, Y.J.; Yue, Y.F.; Qian, G.D.; Chen, B.L. Luminescent functional metal-organic frameworks. *Chem. Rev.* **2012**, *112*, 1126–1162. [CrossRef] [PubMed]
22. Della Rocca, J.; Liu, D.M.; Lin, W.B. Nanoscale metal-organic frameworks for biomedical imaging and drug delivery. *Acc. Chem. Res.* **2011**, *44*, 957–968. [CrossRef] [PubMed]
23. Gu, Z.Y.; Chen, Y.J.; Jiang, J.Q.; Yan, X.P. Metal-organic frameworks for efficient enrichment of peptides with simultaneous exclusion of proteins from complex biological samples. *Chem. Commun.* **2011**, *47*, 4787–4789. [CrossRef] [PubMed]
24. Chen, X.F.; Zang, H.; Wang, X.; Cheng, J.G.; Zhao, R.S.; Cheng, C.G.; Lu, X.Q. Metal-organic framework MIL-53(Al) as a solid-phase microextraction adsorbent for the determination of 16 polycyclic aromatic hydrocarbons in water samples by gas chromatography-tandem mass spectrometry. *Analyst* **2012**, *137*, 5411–5419. [CrossRef] [PubMed]
25. Xie, L.J.; Liu, S.Q.; Han, Z.B.; Jiang, R.F.; Liu, H.; Zhu, F.; Zeng, F.; Su, C.Y.; Ouyang, G.F. Preparation and characterization of metal-organic framework MIL-101(Cr)-coated solid-phase microextraction fiber. *Anal. Chim. Acta* **2015**, *853*, 303–310. [CrossRef] [PubMed]
26. Huo, S.H.; Yan, X.P. Facile magnetization of metal-organic framework MIL-101 for magnetic solid-phase extraction of polycyclic aromatic hydrocarbons in environmental water samples. *Analyst* **2012**, *137*, 3445–3451. [CrossRef] [PubMed]
27. Cui, X.Y.; Gu, Z.Y.; Jiang, D.Q.; Li, Y.; Wang, H.F.; Yan, X.P. In situ hydrothermal growth of metal-organic framework 199 films on stainless steel fibers for solid-phase microextraction of gaseous benzene homologues. *Anal. Chem.* **2009**, *81*, 9771–9777. [CrossRef] [PubMed]
28. Zhang, Z.M.; Huang, Y.C.; Ding, W.W.; Li, G.K. Multilayer interparticle linking hybrid MOF-199 for non-invasive enrichment and analysis of plant hormone ethylene. *Anal. Chem.* **2014**, *86*, 3533–3540. [CrossRef] [PubMed]
29. Ge, D.; Lee, H.K. Water stability of zeolite imidazolate framework 8 and application to porous membrane-protected micro-solid-phase extraction of polycyclic aromatic hydrocarbons from environmental water samples. *J. Chromatogr. A* **2011**, *1218*, 8490–8495. [CrossRef] [PubMed]

30. Chang, N.; Gu, Z.Y.; Wang, H.F.; Yan, X.P. Metal-organic framework-based tandem molecular sieves as a dual platform for selective microextraction and high-resolution gas chromatographic separation of n-alkanes in complex matrixes. *Anal. Chem.* **2011**, *83*, 7094–7101. [CrossRef] [PubMed]

31. Zhou, Q.; Zhu, L.; Xia, X.; Tang, H. The water-resistant zeolite imidazolate framework 67 is a viable solid phase sorbent for fluoroquinolones while efficiently excluding macromolecules. *Microchim. Acta* **2016**, *183*, 1839–1846. [CrossRef]

32. Yu, L.Q.; Yan, X.P. Covalent bonding of zeolitic imidazolate framework-90 to functionalized silica fibers for solid-phase microextraction. *Chem. Commun.* **2013**, *49*, 2142–2144. [CrossRef] [PubMed]

33. Wu, Y.Y.; Yang, C.X.; Yan, X.P. Fabrication of metal-organic framework MIL-88B films on stainless steel fibers for solid-phase microextraction of polychlorinated biphenyls. *J. Chromatogr. A* **2014**, *1334*, 1–8. [CrossRef] [PubMed]

34. He, C.T.; Tian, J.Y.; Liu, S.Y.; Ouyang, G.; Zhang, J.P.; Chen, X.M. A porous coordination framework for highly sensitive and selective solid-phase microextraction of non-polar volatile organic compounds. *Chem. Sci.* **2013**, *4*, 351–356. [CrossRef]

35. Hu, Y.L.; Lian, H.X.; Zhou, L.J.; Li, G.K. In situ solvothermal growth of metal-organic framework-5 supported on porous copper foam for noninvasive sampling of plant volatile sulfides. *Anal. Chem.* **2015**, *87*, 406–412. [CrossRef] [PubMed]

36. Gao, J.; Huang, C.; Lin, Y.; Tong, P.; Zhang, L. In situ solvothermal synthesis of metal-organic framework coated fiber for highly sensitive solid-phase microextraction of polycyclic aromatic hydrocarbons. *J. Chromatogr. A* **2016**, *1436*, 1–8. [CrossRef] [PubMed]

37. Zang, X.; Zhang, X.; Chang, Q.; Li, S.; Wang, C.; Wang, Z. Metal-organic framework UiO-67-coated fiber for the solid-phase microextraction of nitrobenzene compounds from water. *J. Sep. Sci.* **2016**, *39*, 2770–2776. [CrossRef] [PubMed]

38. Wang, Y.; Rui, M.; Lu, G.H. Recent applications of metal-organic frameworks in sample pretreatment. *J. Sep. Sci.* **2018**, *41*, 180–194. [CrossRef] [PubMed]

39. Tang, B.; Zhang, J.H.; Zi, M.; Chen, X.X.; Yuan, L.M. Solid-phase extraction with metal-organic frameworks for the analysis of chiral compounds. *Chirality* **2016**, *28*, 778–783. [CrossRef] [PubMed]

40. Zhou, B.; Silva, N.J.O.; Shi, F.N.; Palacio, F.; Mafra, L.; Rocha, J. Co$^{II}$/Zn$^{II}$-(L-Tyrosine) magnetic metal-organic frameworks. *Eur. J. Inorg. Chem.* **2012**, *32*, 5259–5268. [CrossRef]

**Sample Availability:** Samples of (±)-1,1'-bi-2-naphthol, (*R*)-1,1'-bi-2-naphthol and (*S*)-1,1'-bi-2-naphthol are available from the authors.

*molecules*

MDPI

*Article*

# Influence of Ligand Functionalization of UiO-66-Based Metal-Organic Frameworks When Used as Sorbents in Dispersive Solid-Phase Analytical Microextraction for Different Aqueous Organic Pollutants

Iván Taima-Mancera [1], Priscilla Rocío-Bautista [1], Jorge Pasán [2], Juan H. Ayala [1], Catalina Ruiz-Pérez [2], Ana M. Afonso [1], Ana B. Lago [2,*] and Verónica Pino [1,*]

[1]   Departament of Chemistry (Analytical Division), University of La Laguna, 38206 Tenerife, Spain;
      ivan.taima.13@ull.edu.es (I.T.-M.); procio@ull.edu.es (P.R.-B.); jayala@ull.edu.es (J.H.A.);
      aafonso@ull.edu.es (A.M.A.)
[2]   X-ray and Molecular Materials Lab (MATMOL), Physics Department, University of La Laguna,
      38206 Tenerife, Spain; jpasang@ull.edu.es (J.P.); caruiz@ull.edu.es (C.R.-P.)
*     Correspondence: alagobla@ull.edu.es (A.B.L.); veropino@ull.edu.es (V.P.); Tel.: +34-9223-18990 (V.P.)

Received: 11 October 2018; Accepted: 31 October 2018; Published: 3 November 2018

**Abstract:** Four metal-organic frameworks (MOFs), specifically UiO-66, UiO-66-NH$_2$, UiO-66-NO$_2$, and MIL-53(Al), were synthesized, characterized, and used as sorbents in a dispersive micro-solid phase extraction (D-μSPE) method for the determination of nine pollutants of different nature, including drugs, phenols, polycyclic aromatic hydrocarbons, and personal care products in environmental waters. The D-μSPE method, using these MOFs as sorbents and in combination with high-performance liquid chromatography (HPLC) and diode-array detection (DAD), was optimized. The optimization study pointed out to UiO-66-NO$_2$ as the best MOF to use in the multi-component determination. Furthermore, the utilization of isoreticular MOFs based on UiO-66 with the same topology but different functional groups, and MIL-53(Al) to compare with, allowed us for the first time to evaluate the influence of such functionalization of the ligand with regards to the efficiency of the D-μSPE-HPLC-DAD method. Optimum conditions included: 20 mg of UiO-66-NO$_2$ MOF in 20 mL of the aqueous sample, 3 min of agitation by vortex and 5 min of centrifugation, followed by the use of only 500 μL of acetonitrile as desorption solvent (once the MOF containing analytes was separated), 5 min of vortex and 5 min of centrifugation. The validation of the D-μSPE-HPLC-DAD method showed limits of detection down to 1.5 ng·L$^{-1}$, average relative recoveries of 107% for a spiked level of 1.50 μg·L$^{-1}$, and inter-day precision values with relative standard deviations lower than 14%, for the group of pollutants considered.

**Keywords:** metal-organic frameworks; dispersive solid-phase extraction; organic pollutants; analyte partitioning

---

## 1. Introduction

Micro- and mesoporous materials are widespread used for separation and purification purposes due to their excellent adsorption properties. Among these porous materials, metal-organic frameworks (MOFs) have received much attention in the last years due to their unique properties: ordered porous structures, the highest surface areas known and even the possibility of tuning their physiochemical behavior [1]. MOFs are three dimensional porous hybrid materials composed by two main building blocks, metal ions as nodes or connectors and organic molecules as linkers, and the combination of these building units offers a limitless number of possible structures [2]. Thus, MOFs have been included as fashion materials in a wide variety of applications, such as gas storage, catalysis, drug delivery or luminescence sensing, among others [3]. Even so, a great number of concerns still exist surrounding stability and synthesis of MOFs [4], despite the fact that nowadays many MOFs have shown excellent stability under harsh conditions [5] and were able to be synthesized on large scales [6].

The use of MOFs as advanced porous materials for more effective and efficient capture of pollutants from different environmental media is increasing in recent years [7]. An important feature of MOFs, which offers a method to satisfy the adsorbent selection criteria, is the modular nature of the organic linker. The design and modification of MOFs at the molecular level can be achieved generally functionalizing the pore surface. This concept of precise control at the molecular level is probably the most distinguishing characteristic of MOFs as compared to other sorbent materials because they can offer additional adsorption sites and also improve the selectivity of pristine MOFs [8].

The well-known MOF UiO-66 (UiO = University of Oslo (Oslo, Norway)) [9] exhibits exceptional thermal and chemical stability in water and organic solvents, while presenting good adsorption properties [10]. Several studies have already shown the feasibility of functionalizing the UiO-66 material without losing the physicochemical properties of the parent framework, and mainly their advantages in gas capture and gas separation have been analyzed [11]. UiO-66 has shown promising adsorption capacities for organic contaminants such as organic dyes [12] and for inorganic pollutants such as heavy metals [13]. However, the adsorption behavior of emerging pollutants, such as pharmaceutical and personal care products on UiO-66 has been scarcely studied [14], and to the best of our knowledge, there are not studies dealing with functionalized UiO-66 materials in microextraction.

Investigations of MOFs in analytical chemistry are rising [15–18]. Thus, Zhou et al. reported the first example of MOFs used in analytical chemistry [19], using them in an on-line solid phase extraction (SPE) method. Since then, a variety of MOFs coming from the most widely known families have been tested so far in a number of analytical SPE applications [15,16,18,20] and even in chromatography [17]. The dispersive mode of the miniaturized solid-phase extraction method (D-µSPE) is a successful approach widely used in sample preparation given its simplicity [21]. It requires a strong dispersion of the sorbent (in an amount lower than 500 mg) into an aqueous sample containing analytes (i.e., with the aid of vortex or ultrasounds), followed by proper separation of the sorbent containing extracted analytes from the sample, and further elution/desorption of trapped analytes before the chromatographic determination [16,21,22].

It is important mentioning that, while existing an increasing number of studies with MOFs as sorbents in D-µSPE [16], few authors have paid close attention to study the nature of the interactions established between the MOF sorbent and the contaminants. Indeed, the type of interactions that take place during the extraction has not been established completely [14,23,24]. Thus, Rocío-Bautista et al. evaluated the partitioning of target compounds to different MOFs in D-µSPE [24]. The study highlighted the complexity in achieving adequate predictions for the microextraction performance, with successful results mainly linked to the pore environment, pore size, and pore aperture widths of the MOF, together with a clear influence of the metal nature. The nature of the metal of the MOF and its influence in extraction studies have also been quite recently pointed out by Lirio et al. [14], with high importance given to the radius of the metal.

In this sense, the present study evaluates the analytical performance in D-µSPE of UiO-66 and its derivatives UiO-66-NO$_2$ and UiO-66-NH$_2$: with the organic linker functionalized with nitro and amino groups, respectively, but maintaining the isoreticular network. The amino (-NH$_2$) and nitro (-NO$_2$) functional groups were chosen to be representative of polar and hydrophilic functionalities. This linker functionalization produces changes in the physicochemical properties of the frameworks with the purpose of establishing stronger host-guest interactions [25], but it also implies a decrease in surface areas and pore volumes [26]. The MOF MIL-53(Al) is used in this study as a comparative material. The application is devoted to the determination of several water contaminants (of quite different nature) using high-performance liquid chromatography (HPLC) and diode-array detection

(DAD), intending a multi-component determination while trying to give insights on the influence of the ligand functionality on the possible MOF-target contaminant interactions favoring the entire D-μSPE-HPLC-DAD method.

## 2. Experimental

### 2.1. Chemicals, Reagents and Materials

Nine analytes of different nature were studied. Carbamazepine (Cbz, 99.0%), 4-cumylphenol (CuP, 99%), 4-*tert*-octylphenol (*t*-OP, 97%), 4-octylphenol (OP, 99%), benzophenone-3 (BP-3, 99.5%) and chrysene (Chy, 98%) were obtained from Sigma-Aldrich (Steinheim, Germany); and progesterone (Pg, >99.99%) was purchased from US Pharmacopeia Reference Standards (Basel, Switzerland). All these compounds were obtained as solid products. A standard solution containing these compounds was prepared in acetonitrile (ACN) Chromasolv$^{TM}$ liquid chromatography (LC) grade, by Honeywell (Seelze, Germany), at a concentration of 100 mg·L$^{-1}$, and stored at 4 °C. Indeno(1,2,3-cd)pyrene (Ind) and triclosan (Tr) were purchased individually as standard solutions, with a concentration of 10 mg·L$^{-1}$ in acetonitrile (ACN) by Dr. Ehrenstorfer GmbH (Augsburg, Germany). Working standard solutions were prepared by dilution in ultrapure water of these standard solutions, with concentrations dependent on the specific experiment. Table S1 of the Electronic Supplementary Material (ESM) shows several characteristics and the structures of the analytes studied.

Ultrapure water (Milli-Q, ultrapure grade) was obtained by a water purification system A10 MilliPore (Watford, UK). Methanol Chromasolv® (LC grade) was purchased from Sigma-Aldrich. HPLC mobile phases were prepared with ultrapure water and ACN Chromasolv$^{TM}$ LC-MS grade. Both phases were filtered with Durapore® membrane filters of 0.45 μm, supplied by Sigma-Aldrich.

0.2 μm polyvinylidene fluoride (PVDF) syringe filters Whatman$^{TM}$ were purchased from GE Healthcare (Buckinghamshire, UK), and used to filter all eluates and standards before HPLC injection.

Pyrex® centrifuge tubes (Corning Inc., Staffordshire, UK) were used in the microextraction procedure, with dimensions of 10 × 2.6 cm and a volume of 25 mL.

Parr Instrument Company (Moline, IL, USA) supplied Teflon solvothermal reactors and stainless steel autoclaves, which were used in the synthesis of the MOFs.

Zirconium chloride (ZrCl$_4$, 98%), aluminum(III) nitrate nonahydrate (Al(NO$_3$)$_3$·9H$_2$O, >99.99%), HCl (37%, $v/v$), 1,4-benzenedicarboxylic acid (H$_2$BDC, 98%), 2-amino-1,4-benzenedicarboxylic acid (NH$_2$–H$_2$BDC, 99%) and 2-nitro-1,4-dicarboxylic acid (NO$_2$–H$_2$BDC, ≥99%) were purchased from Sigma-Aldrich and used in the synthesis of MOFs. The solvents used in the synthesis and washing of MOFs include: dimethylformamide (DMF, ≥99.5%), acquired to Merck KGaA (Darmstadt, Germany), and methanol (≥99.8%) purchased from PanReac AppliChem (Barcelona, Spain).

Tap water was taken at the laboratory. Two wastewater samples were supplied by an environmental monitoring laboratory. They were sampled in different areas of Tenerife Island (Canary Islands, Spain) using amber glass recipients properly cleaned, avoiding the formation of bubbles during sampling. They were kept in fridge until reaching the laboratory, and then they were filtered through 0.45 μm filters and kept in the dark at 4 °C until analysis.

### 2.2. Synthesis of MOFs

The MOFs used in this study were synthesized according to the procedure reported by Katz et al. [27]. A standard upscale synthesis of UiO-66 was performed by dissolving 233 mg of ZrCl$_4$ (1 mmol) and 246 mg of H$_2$BDC (1.5 mmol) in 15 mL of DMF and 1 mL of concentrated HCl. The resulting mixture was heated in a solvothermal reactor at 150 °C for 24 h. After the solution was cooled to room temperature, the resulting solid was filtered and repeatedly washed with DMF, methanol, and heated at 150 °C for 24 h in order to remove guest molecules from the pores of the crystalline structure. UiO-66-NH$_2$ and UiO-66-NO$_2$ MOFs were synthesized analogously by replacing H$_2$BDC with the equivalent molar amounts of NH$_2$–H$_2$BDC and NO$_2$–H$_2$BDC, respectively.

MIL-53(Al) was prepared according to Loiseau et al. [28]. Briefly, 288 mg of $H_2BDC$ (1.7 mmol) and 1.3 g of $Al(NO_3)_3 \cdot 9H_2O$ (3.5 mmol) were mixed in a solvothermal reactor using 15 mL of ultrapure water. The solution was then heated at 220 °C for 3 days. Afterwards, the autoclave was cooled down to room temperature and the obtained white product was isolated by filtration, washed with water, and air-dried at 50 °C. The MOF was finally heated at 400 °C during 16 h for the activation.

All synthetic conditions and obtained yields for the MOFs are summarized in Table 1.

**Table 1.** Synthetic conditions of MOFs and yields obtained.

| MOF | Structure (Detailed Functionalization for UiO-66) | Metal (mg) | Ligand (mg) | Solvent (mL) | Modulator/mL | Yield (%) |
|---|---|---|---|---|---|---|
| UiO-66 | | $Zr^{4+}$ (233) | terephthalic acid (246) | DMF (15) | HCl (37%, *v*/*v*)/1 | 95 |
| UiO-66-NH₂ | | $Zr^{4+}$ (233) | 2-aminoterephthalic acid (271) | DMF (15) | HCl (37%, *v*/*v*)/1 | 78 |
| UiO-66-NO₂ | | $Zr^{4+}$ (233) | 2-nitroterephthalic acid (317) | DMF (15) | HCl (37%, *v*/*v*)/1 | 97 |
| MIL-53(Al) | | $Al^{3+}$ (1300) | terephthalic acid (288) | $H_2O$ (15) | - | 45 |

DMF: dimethylformamide.

## 2.3. Instruments and Equipment

The HPLC used in the determination was a 1260 Infinity model purchased from Agilent Technologies (Santa Clara, CA, USA), in combination with a DAD 1260 Infinity model also from Agilent Technologies. The quantification wavelengths of the DAD were set at 240 nm for Pg, 254 nm for Ind, 270 nm for Chy, 280 nm for CuP, *t*-OP and OP, and 289 nm for Cbz, BP-3 and Tr. The HPLC system includes a Rheodyne injection valve with an injection loop of 20 μL, supplied by Supelco (Bellefonte, PA, USA). The separation of target analytes was carried out in an ACE Ultra Core 5 SuperC18 (5 μm, 150 × 4.6 mm) analytical column, obtained from Symta (Madrid, Spain), with a safeguard column Pelliguard LC-18 purchased to Supelco. ACN and ultrapure water were employed as mobile phases using a linear gradient at a constant flow rate of 1 mL·min$^{-1}$. The chromatographic method starts at 50% (*v*/*v*) of ACN, keeping it constant for 5 min, then increased up to 80% (*v*/*v*) in 2 min, then increased up to 83% ($v/v$) in the next 2.5 min, and finally reaching 100% ($v/v$) of ACN in the next 3.5 min.

A vortexer from Reax-Control Heidolph™ GmbH (Schwabach, Germany) and a centrifuge model 5720 Eppendorf™ (Eppendorf, Hamburg, Germany) were utilized in the D-µSPE procedure.

The synthesis of the MOFs was carried out in a Universal model UF30 oven supplied by Memmert (Schwabach, Germany).

Phase identification of all MOFs was carried out by X-ray powder diffraction. A X'Pert Diffractometer supplied by PANalytical (Eindhoven, The Netherlands) and operating with Bragg-Brentano geometry was used. Data collection was carried out using Cu K$_1$ radiation ($\lambda = 1.5418$ Å) over the angular range from 5.01° to 80.00° (0.02° steps) with a total exposure time of 30 min.

An Affinity-1 Fourier transform-infrared (FTIR) spectroscope from Shimadzu (Kyoto, Japan) was used in the identification of the functional groups incorporated to UiO-66.

The nitrogen adsorption isotherms were measured on a Gemini V2365 Model, supplied by Micromeritics (Norcross, GA, USA), surface area analyzer at 77 K in the range $0.02 \leq P/P_0 \leq 1.00$. The Brunauer, Emmet and Teller (BET) method was used to calculate the surface area.

Particle sizes of the crystals were determined at 25 °C by dynamic light scattering (DLS) using the Zetasizer equipment from Malvern Instruments (Malvern, UK), with the Zetasizer software v. 7.03.

### 2.4. Dispersive Miniaturized Solid-Phase Extraction Procedure (D-µSPE)

All conditions of the D-µSPE method using all studied MOFs as sorbents were optimized, including the conditions of both: the extraction and the desorption steps. Under optimum conditions, the extraction is carried out adding 20 mg of UiO-66-NO$_2$ over 20 mL of water sample (or aqueous standard, depending on the experiment) in a 25 mL Pyrex® centrifuge tube. The tube is then subjected to vortex stirring for 3 min, to increase the strength of the interaction between the sorbent and the analytes. Then, the phases are separated by centrifugation ($1921\times g$ during 5 min) and the supernatant aqueous phase is carefully removed. For the desorption step, 500 µL of ACN are added to the UiO-66-NO$_2$ left in the tube containing extracted analytes. Vortexing is applied for another 5 min followed by centrifugation during 5 min at $1921\times g$. The eluate is filtered through 0.2 µm PVDF syringe filters before being injected in the HPLC.

## 3. Results and Discussion

### 3.1. Chromatographic Method

Contaminants selected in this study include polycyclic aromatic hydrocarbons, drugs, phenols, and personal care products, with the purpose of having a variety of quite different analytes, thus covering different possible interaction mechanisms with the MOFs. The determination of the nine target compounds was carried out using HPLC-DAD, employing proper quantification wavelengths for each analyte. The optimum conditions for the separation were summarized in Section 2.3, with the overall separation requiring less than 13 min, as it can be observed in Figure S1 of the ESM.

Several quality analytical parameters of the calibrations obtained by HPLC-DAD are shown in Table S2 of the ESM. Calibration curves present adequate linearity, with correlation coefficient (R) values higher than 0.9983. The limits of detection (LOD) and the limits of quantification (LOQ) were calculated as the signal to noise (S/N) ratio of 3 and 10, respectively. LOD and LOQ values were verified by preparation of standards at such levels of concentration. LODs varied from 0.02 µg·L$^{-1}$ for Chy to 1.00 µg·L$^{-1}$ for CuP. The precision of the chromatographic method was calculated using three standards at concentration levels not utilized in the calibration curve (but included within the calibration range): 6 µg·L$^{-1}$, 30 µg·L$^{-1}$ and 70 µg·L$^{-1}$ ($n = 5$). The obtained results are included in Table S3 of the ESM. In all cases, relative standard deviations (RSD, in %) values were lower than 3.7% for the lowest concentration level tested and 2.9% for the highest concentration level injected.

Regarding the precision of the chromatographic retention times, RSD values were always lower than 0.21% (*n* = 15).

## 3.2. Synthesis and Characterization of Studied MOFs

Following an isoreticular synthesis, a family of MOFs based on the UiO-66 structure was obtained from the two different linker ligands: $NH_2-H_2BDC$ and $NO_2-H_2BDC$ [25]. The crystal and particle size of the UiO-66 and UiO-66-X (X = $NH_2$ and $NO_2$) materials were controlled by modulated synthesis with HCl (as detailed in Section 2.2 and Table 1). All compounds were obtained in high yields without loss of crystallinity or porosity by including HCl in the reaction mixtures during the synthesis.

The X-ray diffraction patterns obtained for the as-synthesized samples (see Figures S2 and S3 of the ESM) revealed that the materials are crystalline and the two functionalized compounds are isostructural with the parent material UiO-66, which demonstrates that the tagged UiO-66-X MOFs are topologically equivalent with UiO-66. It is important to proof this equivalence in order to achieve proper comparison when using these materials as sorbents, to clearly link results to the organic ligand nature.

$N_2$ adsorption/desorption isotherms were collected at 77 K (Figure S4 of the ESM) and the Brunauer, Emmett and Teller (BET) surface areas were calculated and all the materials were found to retain porosity. The BET surface area data was found to decrease in surface area with the functionalization of the pores, from ~1342 $m^2 \cdot g^{-1}$ in the parent UiO-66 to ~794 $m^2 \cdot g^{-1}$ for UiO-66-$NH_2$ and ~771 $m^2 \cdot g^{-1}$ for UiO-66-$NO_2$, and they are in agreement with previous reported values [27].

The presence of the functional groups on the linkers was further evidenced by characterizing the MOFs with FTIR spectroscopy (Figure S5 of the ESM). Thus, UiO-66-$NH_2$ displays a broad absorption band at 3336 $cm^{-1}$ that is assigned to the N-H stretching modes. A band at 1546 $cm^{-1}$ and a band at 1389 $cm^{-1}$ are attributed respectively, to the asymmetric ($\nu(NO)_{asym}$) and symmetric ($\nu(NO)_{sym}$) stretching modes of the nitro group in UiO-66-$NO_2$.

MIL-53(Al) was used in this study for comparative purposes, because it has previously demonstrated its successful performance as sorbent in D-µSPE. The diffraction pattern (Figure S3 of the ESM) shows that the synthesized compound presents the crystal structure of MIL-53(Al) and the calculated Brunauer, Emmett and Teller (BET) surface area is in agreement with the reported value [24].

The particle size distribution of the studied MOFs, shown in Figure S6 of the ESM, shows certain dispersion, particularly for MIL-53(Al)—from 0.1 to 1.5 µm—and for UiO-66—from 0.1 to 1.1 µm, with narrower distribution for UiO-66-$NO_2$ and UiO-66-$NH_2$. In general, most crystals have particles sizes ranging between ~0.4–0.5 µm and ~0.7–0.8 µm, thus showing quite similar values.

## 3.3. Screening of MOFs as Sorbents in D-µSPE-HPLC-DAD

D-µSPE was selected as microextraction approach in this study given its simplicity and high analytical performance [16]. An initial screening study was carried out in order to study the microextraction performance of UiO-66, UiO-66-$NH_2$ and UiO-66-$NO_2$ in the D-µSPE of target contaminants. The main difference among these MOFs, as mentioned above in the characterization study, is the nature of the organic ligand, having all the same topology and quite similar surface areas. In this screening study, the MOF MIL-53(Al) was also included in the comparison for having been pointed out as adequate sorbent in common microextraction applications of similar analytes [24].

The D-µSPE method was initially performed with common extraction conditions. Thus, low amounts of MOF were used to fulfil microextraction requirements, minimization of costs, and proper environmental goals. The analytes, once trapped by the MOF, were desorbed using ACN as elution solvent, for being compatible with the HPLC mobile phases used. In this sense, the initial working conditions included 20 mg of MOF, 20 mL of an aqueous standard (containing all analytes at a concentration level of 5.00 µg·L$^{-1}$), 5 min of vortex agitation, and 5 min of centrifugation (at 1921× *g*). Afterwards, the supernatant was discarded and 0.500 mL (to avoid losses of preconcentration during the process) of ACN were added as elution solvent, followed by 5 min of vortex and 5 min of centrifugation (at 1921× *g*).

Results obtained under these conditions revealed that UiO-66-NO$_2$ presented adequate extraction efficiencies for many analytes, and that in practically all cases (only excluding CuP), MIL-53(Al) was not the best MOF to be chosen as sorbent despite previous studies [24]. Therefore, we used the information obtained from this screening study to optimize the entire D-µSPE-HPLC-DAD method using UiO-66-NO$_2$ as sorbent.

### 3.4. Optimization of the D-µSPE-HPLC-DAD Method Using UiO-66-NO$_2$

The D-µSPE-HPLC-DAD method with UiO-66-NO$_2$ was optimized having as targets the maximization of the extraction efficiency for the highest number of possible of contaminants, and the minimization of the amount of MOF and solvents in the procedure. Main variables studied in the one-factor-at a time optimization of the method were: amount of MOF sorbent, extraction and elution times, number of elution steps and nature of the elution solvent.

The first variable optimized was the amount of UiO-66-NO$_2$, with tested values ranging between 10 and 30 mg. These studies were carried out with aqueous standards (10.0 µg·L$^{-1}$ of Cbz, Tr and *t*-OP, and 2.50 µg·L$^{-1}$ for the rest of analytes), and the remaining conditions already fixed during the screening (Section 3.3). Figure 1 shows the results obtained monitoring the extraction efficiencies in terms of peak areas. In general, best results were obtained when using 10 or 20 mg of UiO-66-NO$_2$, without significant differences in the performance among them, except for Pg, which achieved much better results when using 20 mg. For this reason, 20 mg was selected as the optimum amount of UiO-66-NO$_2$.

The second variable considered in this optimization was the extraction time, with the extraction step assisted by vortex, utilizing times between 1 and 5 min. Vortex times higher than 5 min are not advisable (unhealthy) and thus were not tested. These experiments were performed using 20 mg of MOF, and the fixed conditions above mentioned. Figure S7 of the ESM shows the obtained results, which clearly point out to 3 min as the optimum value.

The third variable included in the optimization was the elution time, also assisted by vortex, with times also varying between 1 and 5 min, using as fixed conditions: 20 mg of MOF, 3 min for the vortex-assisted extraction time, and the remaining conditions fixed as in previous experiments. Figure S8 of the ESM shows clear improvements in the elution with longer times, and thus 5 min was selected as the optimum.

The elution step in D-µSPE with MOFs has been pointed out as the critical step to achieve adequate analytical performance [29]. Therefore, the number of elution steps and the nature of the elution solvent were also considered in this optimization. Regarding the number of elution steps, it was evaluated the use of two elution steps (each one with 0.250 mL of ACN) versus the use of one single elution step (0.500 mL of ACN). It can be observed from Figure S9 of the ESM the absence of significant improvements when increasing the number of elution steps. Therefore, one single elution step was preferred. This also permits a decrease in the overall analysis time, which is advisable. Regarding the nature of the elution solvent, different solvents (compatible with HPLC mobile phases) were compared: ACN, methanol, and acetone; using all already optimized conditions of the method. Results (included in Figure S10 of the ESM) point out the adequacy of ACN. A summary of the entire optimized D-µSPE-HPLC-DAD method is included in Figure 2.

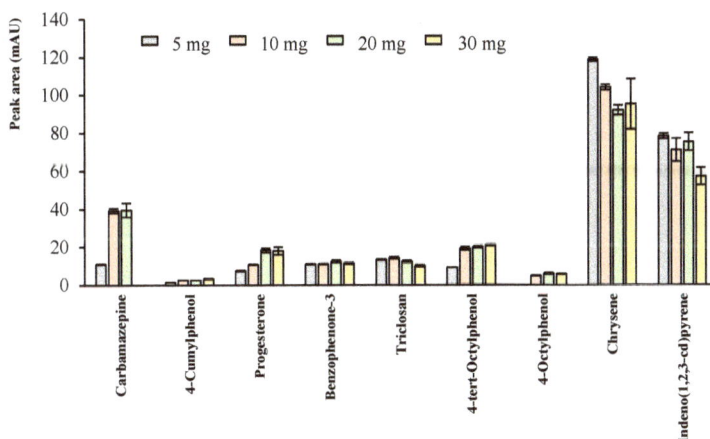

**Figure 1.** Effect of the amount of UiO-66-NO$_2$ on the extraction efficiency for all analytes in D-µSPE-HPLC-DAD. Specific conditions are described in Section 3.4. Experiments were carried out in triplicate.

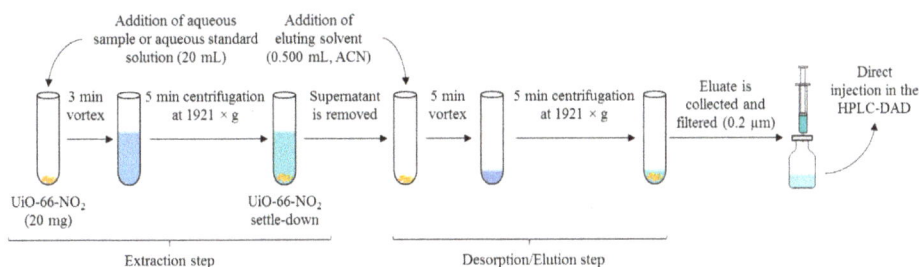

**Figure 2.** Scheme of the entire D-µSPE-HPLC-DAD method using the MOF UiO-66-NO$_2$ under optimum conditions.

## 3.5. Influence of the UiO-66 Ligand Functionalization in the Overall Efficiency of the D-µSPE-HPLC-DAD Method

After completion of the optimization, the D-µSPE-HPLC-DAD method was carried out using UiO-66, UiO-66 derivatives, and MIL-53(Al) as sorbents. The $E_R$ values obtained for the target analytes with the different MOFs are included in Figure 3. Results revealed that UiO-66-NO$_2$ was the best MOF as sorbent for seven out of the nine contaminants studied, and thus it can be considered a generic sorbent if intending a multi-component determination.

Nevertheless, apart for the importance of utilizing this MOF to set up a multi-component monitoring method through D-µSPE, the obtained results are highly valuable in order to get understanding on the nature of possible interactions MOFs-analytes. However, one should keep in mind that the $E_R$ values refer to the complete microextraction method, comprising two steps: the adsorption of the analytes by the MOF extractant and the elution/desorption process. Therefore, the affinity of the MOF for a particular analyte cannot be made exclusively based on the $E_R$ value, since this evaluate a complete process in which desorption readiness (weak analyte-MOF interaction) is favored.

In general, the results introduced in this study indicate that the functionalization by means of polar groups is a noteworthy factor affecting positively the total efficiency of the method (understood as extraction/elution process ability) for the studied analytes of quite different nature. This result shows that the reduction in effective pore size by functional groups is not critical in the analytical procedure, since the electronic properties of these groups favor the analyte-MOF interaction. The

UiO-66 has a moderate pore window size of around 7 Å in diameter and moderate pore size diameter of 11 Å, whereas the UiO-66-X variants reduce the window opening and pore sizes [30]. This implies that the pore window size is large enough for small molecules, but the bulkier ones may have hindering problems and they most likely interact with the MOF surface.

Two adsorption sites for pollutants molecules can be distinguished in the UiO-66-X family, the $Zr_6O_4(OH)_4$ cluster, a hydrophilic area where water, acetic acid and other groups can be coordinated in some MOFs [31]; and the ligand environment which changes from the more hydrophobic X = H to the more hydrophilic X = $NH_2$ or $NO_2$. The adsorption mechanisms may include hydrophobic effects, $\pi$–$\pi$ electron donor–acceptor interactions, electrostatic attractions, [32] or even stronger interactions such as chemical bond or hydrogen bond [33,34].

Analytes accessing the pores window will have a strong interaction with the inside pore walls, which probably improve the extraction step but at the same time, makes more difficult the elution step. The host voids in functionalized materials can act as a "tweezer", providing suitable electronic environments to trap the guest molecules [8]. The moderate pore size will make the adsorbed molecules have a closer distance and stronger interactions with the inside pore walls, thus making difficult the elution process with organic eluents. Therefore, we cannot conclude that higher real recovery values listed in Figure 3 necessarily imply better analyte-MOF interactions, they just represent the minimum value of the quantity of analyte adsorbed by the MOF (either at the pores or at the surface).

Considering this, the best extraction performance occurs for the 4-octylphenol with both $NH_2$- and $NO_2$-functionalized UiO-66 with $E_R$ values larger than 60%, and in general, UiO-66-$NO_2$ seems to be the best extractant material as remarked before. The influence of the narrower pore in the decorated UiO-66 MOFs can be observed in the trend for the three phenols analyzed. They follow a trend where the 4-cumylphenol (CuP) is the bulkier, followed by the 4-*tert*-octylphenol (*t*-OP) and the 4-octylphenol (OP). The $E_R$ values for the $NH_2$-UiO-66 and $NO_2$-UiO-66 follow an inverse trend, OP > *t*-OP > CuP, whilst for the bare UiO-66 and MIL-53(Al) the $E_R$ values are similar (around 40%). One can conclude that for bulkier analytes the adsorption is hindered, and the $E_R$ is reduced. In the case of the MOF MIL-53(Al), its breathing nature implies that the structure can change to the closed-form upon adsorption of guests, in particular, water or some analytes can be triggering this transformation and the resulting analytical performance is affected.

In general, the functionalized $NH_2$- and $NO_2$-UiO-66 MOFs outperform the bare UiO-66 in recovery values. This may be due to various reasons: (i) The amino -and nitro- decorations as H-bond donor and acceptor groups increase the anchoring sites for guest molecules, and these electron-withdrawing groups could lead to effective adsorption from charge-transfer interactions between the functionalized groups and these guest molecules [35]; (ii) The more hydrophilic environment in the pore surface caused by the $NH_2$ and $NO_2$ groups may promote a better elution with the non-polar organic solvents.

In the case of triclosan (Tr), it seems that a larger number of anchoring sites in the form of H-bond acceptors or donors notably increase the extraction ability of the UiO-66 [36]. This situation is also observed in Figure 3, where both $NH_2$- and $NO_2$-UiO-66 outperform bare UiO-66 in the extraction of triclosan, and a similar trend is observed for carbamazepine (Cbz), progesterone (Pg), and benzophenone-3 (BP-3).

In the case of the PAHs included in this analysis, chrysene (Chy) and indeno(1,2,3-cd)pyrene (Ind), the functionalized UiO-66 MOFs also perform better than the bare one. This situation seems contradictory, since more hydrophobic, bulky analytes are being better recovered by more hydrophilic, narrow pore $NH_2$- and $NO_2$-UiO-66 materials. This behavior may be explained taking into account the favored elution process of loosely linked PAHs on the surface of the functionalized MOFs. Probably, these bulky PAHs are adsorbed on the surface of the MOFs and the slightly more hydrophilic environment of the functionalized UiO-66 help in the elution process, and we therefore observe an increase in the total $E_R$ factor. Also, we cannot discard some competitive interactions of the MOFs with the other analytes that lead to a better recovery of the PAHs in this case.

Clearly, to establish relationships between the effects of the functionalization of MOFs on their sorption capacities results a very difficult task, because there are numerous factors to consider [37]. Among them, pore size reduction and molecular sieving effect [38], changes in polar character or electronic environments of the frameworks, energy effects [39] or other factors (such as intrinsic defects, co-adsorption, or diffusive transport in the material pores, as those most remarkable). Moreover, from the point of view of organic pollutants trapped by these sorbents, the physicochemical parameters such as molecular sizes, shapes, polarities, polarizabilities, interaction abilities, solubility, hydrophilicity, acidity and so on, also need consideration. An approach to the fundamental understanding of mechanisms and interactions is highly required for the design of better materials, even selective, for microextraction processes.

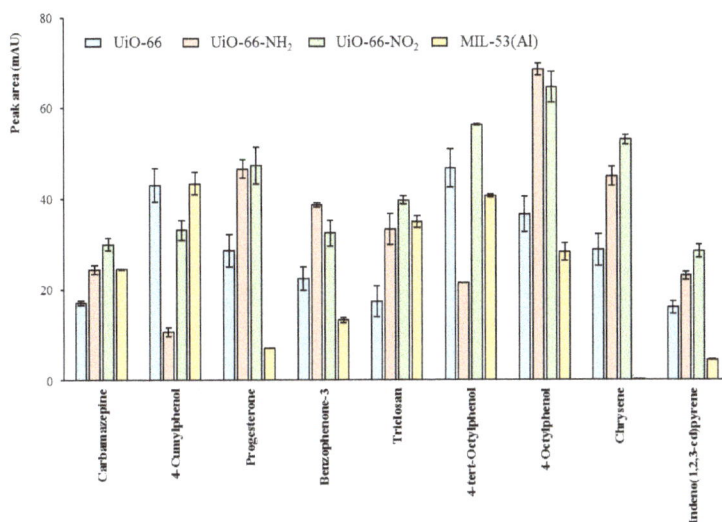

**Figure 3.** Comparison between the different MOFs used in terms of extraction efficiency with the optimum D-µSPE-HPLC-DAD method. Fixed optimum conditions as described in the text. Experiments were carried out in triplicate.

### 3.6. Quality Analytical Parameters of the Optimized D-µSPE-HPLC-DAD Method

The analytical method using D-µSPE-HPLC-DAD and the MOF UiO-66-NO$_2$ under the optimized conditions described (Figure 2) was validated. Calibration curves were obtained subjecting aqueous standards to the entire method. Several of the quality analytical parameters obtained are shown in Table 2. Correlation coefficients (R) were higher than 0.9966 in all cases. LODs were calculated as the concentration in the aqueous sample able to generate a signal to noise ratio of three (S/N = 3) after the overall microextraction and chromatographic procedure, and LOQs as ten times signal to noise ratio (S/N = 10). LODs and LOQs were verified by preparation of aqueous standards at those levels. Obtained LODs ranged from 1.5 ng·L$^{-1}$ for Chy and Ind to 300 ng·L$^{-1}$ for CuP and *t*-OP. It is important to highlight the low LODs achieved with the current microextraction method, particularly considering that DAD is used. A comparison with other methods, which also use solid-based dispersion techniques in combination with HPLC-DAD or HPLC-UV, are included in Table S4 of the ESM. Clearly, the current study presents better sensitivity, and it presents the difficulty of dealing with multi-component determination.

The inter-day precision (RSD, in %) of the entire D-µSPE-HPLC-DAD method was studied, by triplicate, at two spiked levels during three non-consecutive days (*n* = 3, intra-day). Table 3 shows that inter-day RSD values ranged from 4.1% for Pg to 14% for Cbz at the lowest spiked level, and from 4.3% for OP to 9.7% for Tr at the highest spiked level. Intra-day RSD values were always lower than 12%.

**Table 2.** Several quality analytical parameters of the D-μSPE-HPLC-DAD method.

| Analyte | Calibration Range (µg·L⁻¹) | R | $s_{y/x}$ [a] | Slope ± SD [b] | LOD (ng·L⁻¹) | LOQ (ng·L⁻¹) |
|---|---|---|---|---|---|---|
| Carbamazepine | 0.05–5.74 | 0.9989 | 0.27 | 2.3 ± 0.2 | 5.0 | 16.7 |
| 4-Cumylphenol | 0.80–5.74 | 0.9966 | 0.11 | 0.5 ± 0.1 | 90 | 300 |
| Progesterone | 0.01–5.74 | 0.9980 | 0.99 | 6.5 ± 0.5 | 2.4 | 8.00 |
| Benzophenone-3 | 0.05–5.74 | 0.9991 | 0.38 | 3.8 ± 0.2 | 4.5 | 15.0 |
| Triclosan | 0.50–5.00 | 0.9982 | 0.17 | 1.3 ± 0.1 | 30 | 100 |
| 4-*tert*-Octylphenol | 0.50–4.00 | 0.9995 | 0.09 | 1.8 ± 0.1 | 90 | 300 |
| 4-Octylphenol | 0.10–5.00 | 0.9998 | 0.10 | 2.4 ± 0.1 | 15 | 50.0 |
| Chrysene | 0.01–5.74 | 0.9984 | 4.8 | 37 ± 2 | 1.5 | 5.00 |
| Indeno(1,2,3-cd)pyrene | 0.01–5.74 | 0.9986 | 1.6 | 13 ± 1 | 1.5 | 5.00 |

[a] standard deviation of the regression (or error of the estimate). [b] confidence intervals for the slope ($n = 6$) with a signification level of 95%.

**Table 3.** Analytical performance of the entire D-μSPE-HPLC-DAD method in terms of relative recovery, extraction efficiency, and inter-day precision with aqueous standards.

| Analyte | Spiked Level 1 (1.50 µg·L⁻¹) | | | | Spiked Level 2 (4.50 µg·L⁻¹) | | | |
|---|---|---|---|---|---|---|---|---|
| | $E_R$ [a] (%) | RR [b] (%) | Inter-Day RSD [c] (%) | Intra-Day RSD Range [d] (%) | $E_R$ [a] (%) | RR [b] (%) | Inter-Day RSD [c] (%) | Intra-Day RSD Range [d] (%) |
| Cbz | 22.0 | 99.4 | 14 | 1.0–12 | 15.6 | 100 | 8.8 | 6.9–11 |
| CuP | 35.2 | 126 | 9.3 | 4.7–8.3 | 21.1 | 100 | 9.6 | 5.4–6.7 |
| Pg | 51.0 | 111 | 4.1 | 3.3–4.4 | 42.8 | 88.8 | 6.7 | 3.2–4.8 |
| BP-3 | 29.2 | 112 | 9.4 | 3.2–7.7 | 25.7 | 91.9 | 8.1 | 2.6–3.6 |
| Tr | 40.8 | 95.0 | 8.2 | 5.4–9.5 | 43.0 | 104 | 9.7 | 5.6–8.9 |
| t-OP | 53.5 | 118 | 7.2 | 4.1–8.2 | 45.7 | 102 | 5.7 | 2.5–3.7 |
| OP | 69.6 | 102 | 7.5 | 6.1–9.5 | 63.9 | 90.5 | 4.3 | 1.2–2.4 |
| Chy | 39.4 | 109 | 5.5 | 2.0–7.4 | 43.8 | 127 | 8.7 | 3.3–9.6 |
| Ind | 27.1 | 87.3 | 9.1 | 4.1–8.4 | 24.3 | 79.2 | 6.3 | 2.9–5.9 |

[a] extraction efficiency calculated considering the preconcentration achieved with the microextraction method. [b] relative recovery. [c] relative standard deviation for the inter-day precision ($n = 9$, 3 non-consecutive days). [d] intra-day relative standard deviation range ($n = 3$).

Average relative recoveries (RR, in %) were of 107% at the lowest spiked level and of 98.0% at the highest spiked level. Real extraction efficiencies ($E_R$, in %), for the entire D-μSPE-HPLC-DAD method range from 22.0% for Cbz to 69.6% for OP at the lowest spiked level. Several authors have pointed out the difficulties in reaching $E_R$ values close to 100% in any microextraction procedure. Indeed, $E_R$ values are generally not reported in microextraction studies (see Table S4 of the ESM). In any case, the validity of any $E_R$ value for an analyte in a specific microextraction method is dependent on how sensitive and reproducible is a method for a particular application [16,40].

### 3.7. Analysis of Wastewaters and Tap Water Samples Using the Optimized D-μSPE-HPLC-DAD Method

One tap water and two wastewaters samples were analyzed using the optimized D-μSPE-HPLC-DAD method. As it can be observed in Table S5 of the ESM, none of the nine analytes were detected in these samples.

Wastewater-1 was used as blank matrix to perform precision and recovery studies that serve, in addition, to evaluate the matrix effect given its complexity. The obtained results are also shown in Table S5 of the ESM, using a quite low spiked level of 1.50 µg·L⁻¹. From the obtained results, there is a clear matrix effect for analytes such as Cbz, Tr and Pg. For this type of samples, matrix-matched calibrations are recommended.

In any case, it is important to mention that the obtained RSD values (in %) were lower than 17%, which is adequate for this microextraction method considering the low spiked level used for the wastewater sample and its complexity.

## 4. Conclusions

Four MOFs (MIL-53(Al), UiO-66, UiO,66-NH$_2$ and UiO-66-NO$_2$) were successfully synthesized, characterized and tested as sorbents for a D-μSPE method that implies monitoring of nine pollutants of different nature: carbamazepine, 4-cumylphenol, progesterone, benzophenone-3, triclosan, 4-*tert*-octylphenol, 4-octylphenol, chrysene and indeno(1,2,3-cd)pyrene.

The dispersive method in combination with HPLC-DAD was properly optimized using the UiO-66-NO$_2$ MOF, selected as the sorbent which offers highest extraction efficiencies for seven out of the nine analytes, thus being a generic sorbent for this multi-component determination. Low amounts of MOF (20 mg), low sample volumes (20 mL), short sample preparation times (8 min for extraction and 10 min for desorption, both using vortex), and the minimization of the organic solvent needed in the elution step (500 μL) were the optimum conditions for this microextraction method. Limits of detection down to 1.5 ng·L$^{-1}$ were achieved for Chy and Ind despite using DAD, as well as proper analytical performance results such as adequate recovery, extraction efficiency and inter-day precision.

An insight on the possible interactions established between these MOFs and the studied analytes, as a function of the nature of the functionalization of the organic ligand in the MOF, while keeping constant the remaining topological conditions of the crystals, was given. The total efficiency of the D-μSPE-HPLC-DAD method was positively influenced by the presence of functionalization groups in the ligands of UiO-66, particularly due to the polar character given to the organic linkers. Nevertheless, not all results can be justified only based on the polar character of the organic linkers. As main factors, it is important to highlight the pore size of a MOF and the molecular sieving effect, changes in the polar character or the electronic environments of the frameworks, energy effects, as well as intrinsic characteristics of analytes experiencing partitioning to the MOFs.

The results of the current study will serve to better design of MOFs to be used as sorbents in D-μSPE, intending tailored microextractions.

**Supplementary Materials:** The supplementary materials are available online.

**Author Contributions:** Conceptualization, A.M.A and A.B.L.; Formal analysis, A.B.L.; Funding acquisition, V.P.; Investigation, J.P. and V.P.; Resources, C.R.-P.; Software, J.H.A.; Supervision, J.H.A.; Validation, I.T.-M. and P.R.-B.; Writing—original draft, I.T.-M. and P.R.-B.; Writing—review & editing, J.P., A.M.A., A.B.L. and V.P.

**Funding:** V.P. and C.R.-P. thank the MINECO for the Projects Ref. MAT2014-57465-R and MAT2017-89207-R. P.R.-B. thanks her FPI PhD research contract associated to the Project Ref. MAT2014-57465-R.

**Acknowledgments:** All authors thank the SEGAI services of the ULL, specially the technicians, for the help provided.

**Conflicts of Interest:** The authors declare no conflict of interest. The funders had no role in the design of the study; in the collection, analyses, or interpretation of data; in the writing of the manuscript, and in the decision to publish the results.

## References

1. Yaghi, O.M.; O'Keeffe, M.; Ockwig, N.W.; Chae, H.K.; Eddaoudi, M.; Kim, J. Reticular synthesis and the design of new materials. *Nature* **2003**, *423*, 705–714. [CrossRef] [PubMed]
2. Tranchemontagne, D.J.; Mendoza-Cortés, J.L.; O'Keeffe, M.; Yaghi, O.M. Secondary building units, nets and bonding in the chemistry of metal-organic frameworks. *Chem. Soc. Rev.* **2009**, *38*, 1257–1283. [CrossRef] [PubMed]
3. Furukawa, H.; Cordova, K.E.; O'Keeffe, M.; Yaghi, O.M. The chemistry and applications of metal-organic frameworks. *Science* **2013**, *341*, 974–986. [CrossRef] [PubMed]
4. Schoenecker, P.M.; Carson, C.G.; Jasuja, H.; Flemming, C.J.J.; Walton, K.S. Effect of water adsorption on retention of structure and surface area of metal-organic frameworks. *Ind. Eng. Chem. Res.* **2012**, *51*, 6513–6519. [CrossRef]
5. Rocío-Bautista, P.; Pino, V.; Ayala, J.H.; Ruiz-Pérez, C.; Vallcorba, O.; Afonso, A.M.; Pasán, J. A green metal-organic framework to monitor water contaminants. *RSC Adv.* **2018**, *8*, 31304–31310. [CrossRef]

6. Wang, C.; Liu, X.; Demir, N.K.; Chen, J.P.; Li, K. Applications of water stable metal-organic frameworks. *Chem. Soc. Rev.* **2016**, *45*, 5107–5134. [CrossRef] [PubMed]

7. Li, J.; Wang, X.; Zhao, G.; Chen, C.; Chai, Z.; Alsaedi, A.; Hayat, T.; Wang, X. Metal-organic framework-based materials: Superior adsorbents for the capture of toxic and radioactive metal ions. *Chem. Soc. Rev.* **2018**, *47*, 2322–2356. [CrossRef] [PubMed]

8. Duerinck, T.; Bueno-Perez, R.; Vermoortele, F.; De Vos, D.E.; Calero, S.; Baron, G.V.; Denayer, J.F.M. Understanding hydrocarbon adsorption in the UiO-66 metal-organic framework: Separation of (un)saturated linear, branched, cyclic adsorbates, including stereoisomers. *J. Phys. Chem.* **2013**, *117*, 12567–12578. [CrossRef]

9. Cavka, J.H.; Jakobsen, S.; Olsbye, U.; Guillou, N.; Lamberti, C.; Bordiga, S.; Lillerud, K.P. A new zirconium inorganic building brick forming metal organic frameworks with exceptional stability. *J. Am. Chem. Soc.* **2008**, *130*, 13850–13851. [CrossRef] [PubMed]

10. Ragon, F.; Campo, B.; Yang, Q.; Martineau, C.; Wiersum, A.D.; Lago, A.; Guillerm, V.; Hemsley, C.; Eubank, J.F.; Vishnuvarthan, M.; et al. Acid-functionalized UiO-66(Zr) MOFs and their evolution after intra-framework cross-linking: Structural features and sorption properties. *J. Mater. Chem. A* **2015**, *3*, 3294–3309. [CrossRef]

11. Yang, Q.; Vaesen, S.; Ragon, F.; Wiersum, A.D.; Wu, D.; Lago, A.; Devic, T.; Martineau, C.; Taulelle, F.; Llewellyn, P.L.; et al. A water stable metal-organic framework with optimal features for $CO_2$ capture. *Angew. Chem.-Int. Edit.* **2013**, *52*, 10316–10320. [CrossRef] [PubMed]

12. He, Q.; Chen, Q.; Lü, M.; Liu, X. Adsorption behavior of rhodamine B on UiO-66. *Chin. J. Chem. Eng.* **2014**, *22*, 1285–1290. [CrossRef]

13. Saleem, H.; Rafique, U.; Davies, R.P. Investigations on post-synthetically modified UiO-66-$NH_2$ for the adsorptive removal of heavy metal ions from aqueous solution. *Microporous Mesoporous Mater.* **2016**, *221*, 238–244. [CrossRef]

14. Lirio, S.; Shih, Y.H.; Hsiao, S.Y.; Chen, J.H.; Chen, H.T.; Liu, W.L.; Lin, C.H.; Huang, H.Y. Monitoring the effect of different metal centers in metal-organic frameworks and their adsorption of aromatic molecules using experimental and simulation studies. *Chem. Eur. J.* **2018**, *24*, 14044–14047. [CrossRef] [PubMed]

15. Rocío-Bautista, P.; Pacheco-Fernández, I.; Pasán, J.; Pino, V. Are metal-organic frameworks able to provide a new generation of solid-phase microextraction coatings?—A review. *Anal. Chim. Acta* **2016**, *939*, 26–41. [CrossRef] [PubMed]

16. Rocío-Bautista, P.; González-Hernández, P.; Pino, V.; Pasán, J.; Afonso, A.M. Metal-organic frameworks as novel sorbents in dispersive-based microextraction approaches. *Trac-Trends Anal. Chem.* **2017**, *90*, 114–134. [CrossRef]

17. Pacheco-Fernández, I.; González-Hernández, P.; Pasán, J.; Ayala, J.H.; Pino, V. The rise of metal-organic frameworks in analytical chemistry. In *Handbook of Smart Materials in Analytical Chemistry*, 1st ed.; De la Guardia, M., Esteve-Turrillas, F.A., Eds.; Wiley: Weinheim, Germany, 2019; Volume 1, pp. 463–502.

18. Maya, F.; Cabello, C.P.; Frizzarin, R.M.; Estela, J.M.; Palomino, G.T.; Cerdà, V. Magnetic solid-phase extraction using metal-organic frameworks (MOFs) and their derived carbons. *Trac-Trends Anal. Chem.* **2017**, *90*, 142–152. [CrossRef]

19. Gu, Z.Y.; Wang, G.; Yan, X.P. MOF-5 metal-organic framework as sorbent for in-field sampling and preconcentration in combination with thermal desorption GC/MS for determination of atmospheric formaldehyde. *Anal. Chem.* **2010**, *82*, 1365–1370. [CrossRef] [PubMed]

20. Wang, Y.; Rui, M.; Lu, G. Recent applications of metal-organic frameworks in sample pretreatment. *J. Sep. Sci.* **2018**, *41*, 180–194. [CrossRef] [PubMed]

21. Socas-Rodríguez, B.; Herrera-Herrera, A.V.; Asensio-Ramos, M.; Hernández-Borges, J. Dispersive solid-phase extraction. In *Analytical Separation Science*, 1st ed.; Anderson, J.L., Berthod, A., Pino, V., Stalcup, A.M., Eds.; Wiley: Weinheim, Germany, 2015; Volume 5, pp. 1525–1569.

22. Płotka-Wasylka, J.; Szczepańska, N.; de la Guardia, M.; Namieśnik, J. Miniaturized solid-phase extraction techniques. *Trac-Trends Anal. Chem.* **2015**, *73*, 19–38. [CrossRef]

23. Chen, C.; Chen, D.; Xie, S.; Quan, H.; Luo, X.; Guo, L. Adsorption behaviors of organic micropollutants on zirconium metal-organic framework UiO-66: Analysis of surface interactions. *ACS Appl. Mater. Interfaces* **2017**, *9*, 41043–41054. [CrossRef] [PubMed]

24.  Rocío-Bautista, P.; Pino, V.; Pasán, J.; López-Hernández, I.; Ayala, J.H.; Ruiz-Pérez, C.; Afonso, A.M. Insights in the analytical performance of neat metal-organic frameworks in the determination of pollutants of different nature from waters using dispersive miniaturized solid-phase extraction and liquid chromatography. *Talanta* **2018**, *179*, 775–783. [CrossRef] [PubMed]

25.  Kandiah, M.; Nilsen, M.H.; Usseglio, S.; Jakobsen, S.; Olsbye, U.; Tilset, M.; Larabi, C.; Quadrelli, E.A.; Bonino, F.; Lillerud, K.P. Synthesis and stability of tagged UiO-66 Zr-MOFs. *Chem. Mater.* **2010**, *22*, 6632–6640. [CrossRef]

26.  Rada, Z.H.; Abid, H.R.; Sun, H.; Shang, J.; Li, J.; He, Y.; Liu, S.; Wang, S. Effects of -$NO_2$ and -$NH_2$ functional groups in mixed-linker Zr-based MOFs on gas adsorption of $CO_2$ and $CH_4$. *Prog. Nat. Sci.* **2018**, *28*, 160–167. [CrossRef]

27.  Katz, M.J.; Brown, Z.J.; Colón, Y.J.; Siu, P.W.; Scheidt, K.A.; Snurr, R.Q.; Hupp, J.T.; Farha, O.K. A facile synthesis of UiO-66, UiO-67 and their derivatives. *Chem. Commun.* **2013**, *49*, 9449–9451. [CrossRef] [PubMed]

28.  Loiseau, T.; Serre, C.; Huguenard, C.; Fink, G.; Taulelle, F.; Henry, M.; Bataille, T.; Férey, G. A rationale for the large breathing of the porous aluminum terephthalate (MIL-53) upon hydration. *Chem. Eur. J.* **2004**, *10*, 1373–1382. [CrossRef] [PubMed]

29.  Rocío-Bautista, P.; Pino, V.; Ayala, J.H.; Pasán, J.; Ruiz-Pérez, C.; Afonso, A.M. The metal-organic framework HKUST-1 as efficient sorbent in a vortex-assisted dispersive micro solid-phase extraction of parabens from environmental waters, cosmetic creams and human urine. *Talanta* **2015**, *139*, 13–20. [CrossRef] [PubMed]

30.  Demir, H.; Walton, K.S.; Sholl, D.S. Computational screening of functionalized UiO-66 materials for selective contaminant removal from air. *J. Phys. Chem. C* **2017**, *121*, 20396–20406. [CrossRef]

31.  Bai, Y.; Dou, Y.; Xie, L.H.; Rutledge, W.; Li, J.R.; Zhou, H.C. Zr-based metal-organic frameworks: Design, synthesis, structure, and applications. *Chem. Soc. Rev.* **2016**, *45*, 2327–2367. [CrossRef] [PubMed]

32.  Lv, G.; Liu, J.; Xiong, Z.; Zhang, Z.; Guan, Z. Selectivity adsorptive mechanism of different nitrophenols on UiO-66 and UiO-66-$NH_2$ in aqueous solution. *J. Chem. Eng. Data* **2016**, *61*, 3868–3876. [CrossRef]

33.  Dias, E.M.; Petit, C. Towards the use of metal-organic frameworks for water reuse: A review of the recent advances in the field of organic pollutants removal and degradation and the next steps in the field. *J. Mater. Chem. A* **2015**, *3*, 22484–22506. [CrossRef]

34.  Hasa, Z.; Jhung, S.H. Removal of hazardous organics from water using metal-organic frameworks (MOFs): Plausible mechanisms for selective adsorptions. *J. Hazard. Mater.* **2015**, *283*, 329–339. [CrossRef] [PubMed]

35.  Karmakar, A.; Samanta, P.; Desai, A.V.; Ghosh, S.K. Guest-responsive metal-organic frameworks as scaffolds for separation and sensing applications. *Acc. Chem. Res.* **2017**, *50*, 2457–2469. [CrossRef] [PubMed]

36.  Song, J.Y.; Ahmed, I.; Seo, P.W.; Jhung, S.H. UiO-66 metal-organic framework with free carboxylic acid: Versatile adsorbents via H-bond for both aqueous and nonaqueous phases. *ACS Appl. Mater. Interfaces* **2016**, *8*, 27394–27402. [CrossRef] [PubMed]

37.  Nandy, A.; Forse, A.C.; Whiterspoon, V.J.; Reimer, J.A. NMR spectroscopy reveals adsorbate binding sites in the metal-organic framework UiO-66(Zr). *J. Phys. Chem. C* **2018**, *122*, 8295–8305. [CrossRef]

38.  Chang, N.; Yan, X.-P. Exploring reverse shape selectivity and molecular sieving effect of metal-organic framework UIO-66 coated capillary column for gas chromatographic separation. *J. Chromatogr. A* **2012**, *1257*, 116–124. [CrossRef] [PubMed]

39.  Ahmed, I.; Jhung, S.H. Applications of metal-organic frameworks in adsorption/separation processes via hydrogen bonding interactions. *Chem. Eng. J.* **2017**, *310*, 197–215. [CrossRef]

40.  Trujillo-Rodríguez, M.J.; Rocío-Bautista, P.; Pino, V.; Afonso, A.M. Ionic liquids in dispersive liquid-liquid microextraction. *Trac-Trends Anal. Chem.* **2013**, *51*, 87–106. [CrossRef]

**Sample Availability:** Not available.

*molecules*

MDPI

*Article*

# Comparison between Exhaustive and Equilibrium Extraction Using Different SPE Sorbents and Sol-Gel Carbowax 20M Coated FPSE Media

Angela Tartaglia [1], Marcello Locatelli [1,*], Abuzar Kabir [2,*], Kenneth G. Furton [2], Daniela Macerola [1], Elena Sperandio [1], Silvia Piccolantonio [1], Halil I. Ulusoy [3], Fabio Maroni [1], Pantaleone Bruni [1], Fausto Croce [1] and Victoria F. Samanidou [4]

[1] Department of Pharmacy, University of Chieti–Pescara "G. d'Annunzio", Via dei Vestini 31, 66100 Chieti, Italy; angela.tartaglia@unich.it (A.T.); daniela.macer@gmail.com (D.M.); sperandioelena94@gmail.com (E.S.); silvia.piccolantonio@studenti.unich.it (S.P.); fabio.maroni@unich.it (F.M.); pantaleonebruni@libero.it (P.B.); fausto.croce@unich.it (F.C.)
[2] Department of Chemistry and Biochemistry, International Forensic Research Institute, Florida International University, 11200 SW 8th St, Miami, FL 33199, USA; furtonk@fiu.edu
[3] Department of Analytical Chemistry, Faculty of Pharmacy, Cumhuriyet University, Sivas 58140, Turkey; hiulusoy@yahoo.com
[4] Laboratory of Analytical Chemistry, Department of Chemistry, Aristotle University of Thessaloniki, 54124 Thessaloniki, Greece; samanidu@chem.auth.gr
* Correspondence: m.locatelli@unich.it (M.L.); akabir@fiu.edu (A.K.); Tel.: +39-0871-3554590 (M.L.); +1-305-348-2396 (A.K.); Fax: +39-0871-3554911 (M.L.); +1-305-348-4172 (A.K.)

Academic Editor: Victoria Samanidou
Received: 13 December 2018; Accepted: 20 January 2019; Published: 22 January 2019

**Abstract:** This paper reports the performance comparison between the exhaustive and equilibrium extraction using classical Avantor C18 solid phase extraction (SPE) sorbent, hydrophilic-lipophilic balance (HLB) SPE sorbent, Sep-Pak C18 SPE sorbent, novel sol-gel Carbowax 20M (sol-gel CW 20M) SPE sorbent, and sol-gel CW 20M coated fabric phase sorptive extraction (FPSE) media for the simultaneous extraction and analysis of three inflammatory bowel disease (IBD) drugs that possess logP values (polarity) ranging from 1.66 for cortisone, 2.30 for ciprofloxacin, and 2.92 for sulfasalazine. Both the commercial SPE phases and in-house synthesized sol-gel CW 20M SPE phases were loaded in SPE cartridges and the extractions were carried out under an exhaustive extraction mode. FPSE was carried out under an equilibrium extraction mode. The drug compounds were resolved using a Luna C18 column (250 mm × 4.6 mm; 5 μm particle size) in gradient elution mode within 20 min and the method was validated in compliance with international guidelines for the bioanalytical method validation. Novel in-house synthesized and loaded sol-gel CW 20M SPE sorbent cartridges were characterized in terms of their extraction capability, breakthrough volume, retention volume, hold-up volume, number of the theoretical plate, and the retention factor.

**Keywords:** FPSE; in-house loaded SPE; HPLC-PDA; method validation; IBD; extraction

## 1. Introduction

Sample preparation and 'clean-up' have the aim of improving the analytical parameters to enhance detectability and to protect the performance of analytical equipment. However, these steps are time consuming, utilize large volumes of toxic and hazardous organic solvents, and are prone to error. For these reasons, novel extraction technologies are continuously being developed to overcome these drawbacks [1] in order to improve the extraction efficiency, reduce the sample handling, and enhance the reproducibility.

Novel extraction strategies focus on improving the quality of extraction, decreasing the use of solvents to become environmentally friendly, reducing the time for extraction, and the associated costs. Solid-phase extraction (SPE) is the most popular sample preparation technique used for extraction, clean up, and pre-concentration of both environmental and biological samples. In addition, SPE is fast, easy, and green due to its reduced consumption of solvent compared to liquid–liquid extraction, its relative inexpensiveness and ability for easy automation as well as shorter analysis time. The primary advantage is that SPE is a non-equilibrium, exhaustive extraction procedure; the problem with using an equilibrium process is that the analyst may never know when equilibrium has been reached, and the equilibrium distribution may necessitate multiple extractions [2].

Fabric phase sorptive extraction (FPSE) is a new generation solvent-free/solvent-minimized microextraction technique developed by Kabir and Furton [3]. FPSE is a simple, rapid, and green sample preparation technique, which not only simplifies the sample preparation by eliminating many unnecessary steps such as filtration, protein precipitation, solvent evaporation, and sample reconstitution, it also substantially reduces the organic solvent consumption [4]. The permeable fabric substrate used to chemically bind sol-gel derived high performance sorbents in FPSE mimics the flow-through extraction mechanism used in SPE even when the extraction is carried out under equilibrium extraction mode, resulting in a faster mass transfer and extraordinarily high analyte recovery. The flexibility of the FPSE media allows its direct insertion into the sample container, reducing possible analyte loss [1] and potential cross contamination stems from the containers. The main advantages of FPSE are its high contact surface area for rapid sorbent-analyte interactions and the possibility of enhancing the diffusion of analytes through the FPSE media with magnetic stirring or sonication to obtain a fast extraction equilibrium [5].

Among different treatment regimens used in inflammatory bowel disease (IBD), cortisone, ciprofloxacin, and/or sulfasalazine are orally administrated among other drugs; these drugs are generally present at trace level concentrations in physiological samples including plasma, urine, and saliva. As such, a preconcentration step is necessary to obtain advantageous signal-to-noise ratios in chromatographic separation and detection [6,7]. The logP values of the target analytes ranging from 1.66 for cortisone, to 2.30 for ciprofloxacin, and 2.92 for sulfasalazine (Table 1), coupled to their widespread range of $pK_a$ values pose a great challenge for their simultaneous extraction and analysis. Particularly these differences could influence the extraction efficiency of the analytes. These differences necessarily guide the selection of possible sample preparation and clean-up procedures that can be applied, and failure to select an appropriate sample preparation technique may lead to low analytical performances during traces analysis.

Both the exhaustive extraction technique (such as solid phase extraction) and equilibrium extraction technique (such as solid phase microextraction, fabric phase sorptive extraction, stir bar sorptive extraction) not only utilize different extraction mechanism, but also use mutually exclusive extraction sorbents. Among SPE sorbents, hydrophilic-lipophilic balance (HLB) and C18 are the most popular, whereas for equilibrium extraction, poly(dimethyl siloxane), poly(ethylene glycol), and polyacrylate are noteworthy. As such, it is important to compare the extraction performances between the two major extraction modes as well as between different sorbent chemistries. Only a few sorbent chemistries are available in both the exhaustive and equilibrium extraction formats. However, our research group has recently introduced a large number of traditional microextraction sorbents including poly(dimethyl siloxane) and poly(ethylene glycol) in solid phase extraction format. In the current study, sol-gel CW 20M SPE sorbent and sol-gel CW 20M coated FPSE media are used for extraction performance evaluation and comparison among other sorbents. These phases were newly synthesized in our laboratory and herein tested in order to compare them in terms of the main analytical parameters (for SPE breakthrough volume ($V_B$), retention volume ($V_R$), hold-up volume ($V_M$), retention factor ($k$), and theoretical plates number ($N$), and for both SPE and FPSE the enrichment factors).

**Table 1.** Molecular structures and other pertinent physiochemical properties of selected inflammatory bowel disease (IBD) drugs and the internal standard.

| Drug Name | CAS No. | Molecular Weight (g/mole) | Molecular Structure | LogP | pKa |
|---|---|---|---|---|---|
| Cortisone | 53-06-5 | 360.44 | | 1.66 | 12.6 |
| Methyl-p-hydroxy benzoate | 99-76-3 | 152.15 | | 1.96 | 8.4 |
| Ciprofloxacin | 85721-33-1 | 331.34 | | 2.30 | 5.76 |
| Sulfasalazine | 599-79-1 | 398.39 | | 2.92 | 3.3 |

To the best of our knowledge, this is the first comprehensive comparison between exhaustive and equilibrium extraction modes utilizing multiple C18 SPE sorbents, HLB sorbent, novel sol-gel CW 20M SPE sorbent, and sol-gel CW 20M coated FPSE membrane for the determination of IBD drugs, possessing logP values (polarity) of 1.66 for cortisone, 2.30 for ciprofloxacin, and 2.92 for sulfasalazine. Due to the difference in their physicochemical properties, especially their polarity, selected IDB drugs pose a challenge for their simultaneous extraction and preconcentration at their trace and ultra-trace level concentrations.

## 2. Results and Discussion

*2.1. Determination of SPE Parameters: Breakthrough Volume, Retention Volume, Hold-Up Volume, Retention Factor, and Theoretical Plate Number for SPE Devices*

To understand the extraction mechanism and to utilize their full potential as the extraction media, it is important to characterize the sorbents available for the extraction and clean-up process. When the solid phase extraction (SPE) procedure is adopted, the main parameters that need to be considered are the breakthrough volume ($V_B$), retention volume ($V_R$), hold-up volume ($V_M$), retention factor ($k$), and the theoretical plates number ($N$).

As reported in the literature [8,9], for the determination of the above parameters, *off-line* SPE can be followed by applying the sigmoid shape breakthrough curve. Particularly, from this model obtained by fitting of the experimental points by Boltzmann's function:

$$Y = A_2 + \frac{A_1 - A_2}{1 + e^{\frac{x - x_0}{dx}}} \tag{1}$$

and be calculated by the breakthrough volume

$$V_B = x_0 + dx \cdot ln\left[\frac{100}{99}\left(1 - \frac{A_1}{A_2}\right) - 1\right] \tag{2}$$

hold-up volume:

$$V_M = x_0 + dx \cdot ln\left[99 - 100 \cdot \frac{A_1}{A_2}\right] \tag{3}$$

The retention volume, defined as the inflection point of the curve, corresponds to the $x_0$ value. The retention factor ($k$) is defined using the following equation:

$$V_M = V_R(1 + k) \tag{4}$$

and the theoretical plates number:

$$N = \frac{V_R(V_R - \sigma_V)}{\sigma_V^2} \tag{5}$$

were $\sigma_V$ was obtained by:

$$V_B = V_R - 2\sigma_V \tag{6}$$

In the current study, following our research on innovative extraction procedures [4,10–13] and devices, we compare four different SPE sorbents (both commercially available and in-house produced) and the innovative fabric phase sorptive extraction (FPSE) media.

Using these equations, it was possible to evaluate the breakthrough volume, retention volume, hold-up volume, retention factor, and theoretical plate number for the herein considered stationary phases loaded on the SPE cartridges. In Figure 1 we reported the experimental curves and the fittings used for the parameters calculation.

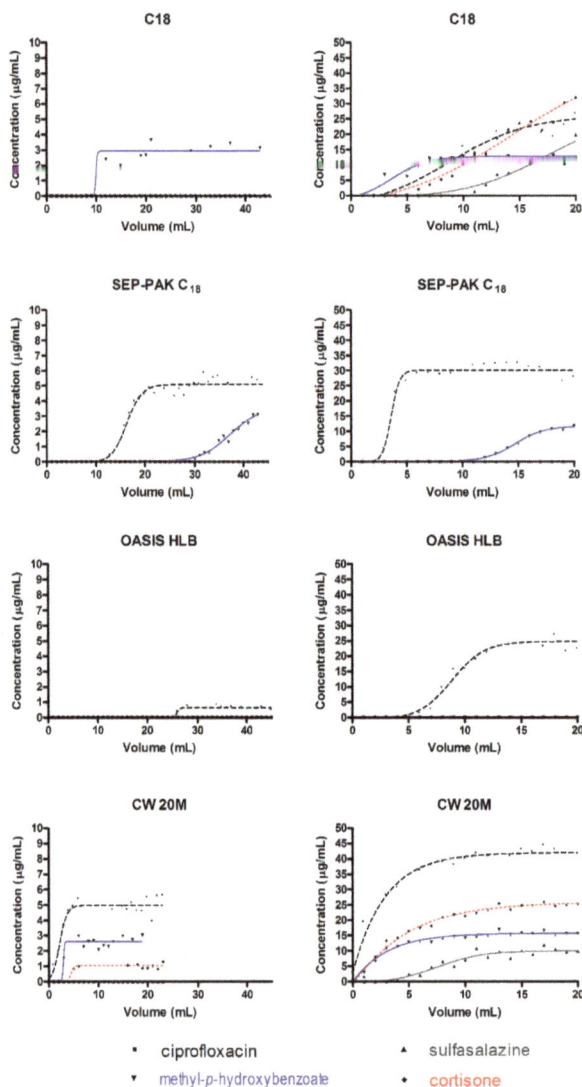

Figure 1. Comparison of breakthrough curves determined for the analytes on the different solid phase extraction (SPE) sorbents and at 10 μg/mL (**left**) and 50 μg/mL (**right**).

The Boltzmann's functions shown in Figure 1 allows us to evaluate the breakthrough volume ($V_B$), retention volume ($V_R$), hold-up volume ($V_M$), retention factor ($k$), and theoretical plate number ($N$) as reported in Tables 2 and 3 (by using the above reported equations). In these Tables, it was highlighted that for the tested compounds, some of these materials are not available in a "universal" phase. In fact, some analytes were not reported (both in Figure 1 and Tables 2 and 3) because no quantitative data were obtained after the SPE steps and, consequently, it was not possible to perform the analysis by Boltzmann's function (like all the other compounds). Similarly, in other published papers [8,9,14], this model could not reach a complete analysis (due to a "floating point error" applying Boltzmann's function) and evaluation of the above-mentioned parameters.

**Table 2.** Parameters determined for the analytes on different sorbents in frontal analysis at the concentration of 10 μg/mL.

| Sorbent | Analyte | Breakthrough Volume (mL) | Retention Volume (mL) | Hold-Up Volume (mL) | Retention Factor (k) | Number of Theoretical Plates (N) |
|---|---|---|---|---|---|---|
| Avantor C$_{18}$ | Ciprofloxacin | >50 | - | - | - | - |
| | Sulfasalazine | >50 | - | - | - | - |
| | Methyl-*p*-hydroxybenzoate | 11.3 | 9.9 | 11.9 | 0.20 | 2000 |
| | Cortisone | >50 | - | - | - | - |
| Sep-Pac Vac C$_{18}$ | Ciprofloxacin | 23.6 | 16.4 | 30.6 | 0.87 | 16 |
| | Sulfasalazine | >50 | - | - | - | - |
| | Methyl-*p*-hydroxybenzoate | 52.8 | 37.3 | 66.8 | 0.79 | 18 |
| | Cortisone | >50 | - | - | - | - |
| Oasis HLB | Ciprofloxacin | 27.6 | 25.9 | 28.8 | 0.11 | 845 |
| | Sulfasalazine | >50 | - | - | - | - |
| | Methyl-*p*-hydroxybenzoate | >50 | - | - | - | - |
| | Cortisone | >50 | - | - | - | - |
| CW 20M | Ciprofloxacin | 5.26 | 2.2 | 8.7 | 2.95 | 1 |
| | Sulfasalazine | >50 | - | - | - | - |
| | Methyl-*p*-hydroxybenzoate | 4.4 | 2.9 | 4.9 | 0.66 | 13 |
| | Cortisone | 6.5 | 4.5 | 7.9 | 0.75 | 17 |

*Molecules* **2019**, 24, 382

**Table 3.** Parameters determined for the analytes on different sorbents in frontal analysis at the concentration of 50 µg/mL.

| Sorbent | Analyte | Breakthrough Volume (mL) | Retention Volume (mL) | Hold-Up Volume (mL) | Retention Factor (k) | Number of Theoretical Plates (N) |
|---|---|---|---|---|---|---|
| Avantor C18 | Ciprofloxacin | 15.1 | 8.8 | 32.2 | 2.68 | 5 |
| | Sulfasalazine | 29.7 | 17.4 | 46.5 | 1.67 | 5 |
| | Methyl-p-hydroxybenzoate | 6.7 | 3.7 | 14.3 | 2.91 | 3 |
| | Cortisone | 25.7 | 14.2 | 49.6 | 2.48 | 4 |
| Sep-Pac Vac C18 | Ciprofloxacin | 5.4 | 3.6 | 7.2 | 0.99 | 12 |
| | Sulfasalazine | >20 | - | - | - | - |
| | Methyl-p-hydroxybenzoate | 21.4 | 14.7 | 27.4 | 0.87 | 15 |
| | Cortisone | >20 | - | - | - | - |
| Oasis HLB | Ciprofloxacin | 13.9 | 8.8 | 20.7 | 1.34 | 8 |
| | Sulfasalazine | >20 | - | - | - | - |
| | Methyl-p-hydroxybenzoate | >20 | - | - | - | - |
| | Cortisone | >20 | - | - | - | - |
| CW 20M | Ciprofloxacin | >20 | - | - | - | - |
| | Sulfasalazine | 13.2 | 7.5 | 22.7 | 2.02 | 4 |
| | Methyl-p-hydroxybenzoate | >20 | - | - | - | - |
| | Cortisone | >20 | - | - | - | - |

All tested SPEs show high retention values for sulfasalazine that do not allow us to evaluate the principal parameters, also due to the fact that with the obtained quantitative data no Boltzmann's function can be performed. Within the four tested phases, the sol-gel CW 20M shows the ability to retain all compounds with a good number of theoretical plates, and in the meantime allow the use of a small elution volume. These findings follow the general principles of the Green Analytical Chemistry. In this case, unfortunately, the calculated breakthrough volume, retention volume, hold-up volume, retention factor, and theoretical plate number were lower than the values obtained for the other tested phases.

Interestingly, commercially available phases show superior results at higher concentration levels for the tested compounds, while the new sol-gel CW 20M are at low concentrations. This finding could be observed from the comparison of Tables 2 and 3 where the evaluated parameters can be calculated for high and low concentrations, respectively.

## 2.2. Mechanism of Extraction in FPSE

FPSE has combined two major sample preparation techniques: SPE (governed by exhaustive extraction principle) and SPME (governed by equilibrium driven extraction principle) into a single sample preparation technique. Due to the geometrical advantage and the combination of both the equilibrium and exhaustive extraction mechanism [15], FPSE can perform exhaustive extraction under equilibrium extraction conditions. The fabric substrates are inherently permeable. The permeability of the fabric substrate remains intact even after the sol-gel sorbent coating. When the FPSE media is immersed into the liquid sample matrix, the aqueous solution rapidly permeates through the porous bed of sol-gel sorbent-coated FPSE media and rapidly interacts with the sorbent via different intermolecular interaction mechanisms. As such, the FPSE membrane behaves like a SPE disk. At the same time, when the FPSE media is introduced into the aqueous solution, it behaves similarly to the SPME thin film format, analytes continue to accumulate onto the FPSE media based on the partition coefficient of the analyte(s) between the sol-gel sorbent and the sample matrix until the extraction equilibrium is reached.

## 2.3. Characterization of Sol-Gel CW 20M SPE Sorbent Using FTIR and TGA Analysis

The experimental FTIR spectrum shows a very weak band at 2974 cm$^{-1}$, indicating the symmetric stretching of a C-H bond. Then a medium sharp band a 1269 cm$^{-1}$, which can be assigned to a Si-CH$_3$ bond and is indicative of the successful integration of methyl trimethoxysilane (MTMS) into the sol-gel network. Then a broad band with a visible shoulder, centered at 1044 cm$^{-1}$, indicative of vibrations related to Si-O-Si bonds. These features are likely connected to the siloxane nature of the sample [16]. All these findings are reported in Figure 2. One very important point that can be noted here, is the absence of any band around 3690 cm$^{-1}$ that corresponds to free silanol (Si-OH). This indicates that even though the sol-gel CW 20M SPE sorbent was not treated with any endcapping (post synthesis derivatization to minimize unreacted surface silanol groups of the silica substrate), the novel SPE sorbent does not possess any residual surface silanol groups.

Thermogravimetric analysis shows a 2% weight loss up to 250 °C, probably due to loss of moisture and volatile species entrapped inside the sol-gel CW20M particles. Then, as indicated by differential thermal analysis (DTA) data, between 400 °C and 600 °C a slightly exothermic process is visible, probably due to the loss of unreacted reaction products trapped inside the sol-gel particles. The sol-gel CW 20M sorbent was synthesized using two catalysts, an acid catalyst during hydrolysis and a base catalyst during polycondensation. Due to the use of two catalysts in the sol-gel synthesis, the CW 20M polymer becomes chemically integrated into the sol-gel network and was able to endure the high temperature without converting into CO$_2$ during the thermogravimetric analysis. It should also be noted that the stoichiometric carbon loading of sol-gel CW 20M was approximately 35% compared to 9–10% carbon loading in commercial C18 sorbents. The total weight loss at the end of the experiment was 6.99%. All these findings are reported in Figure 3.

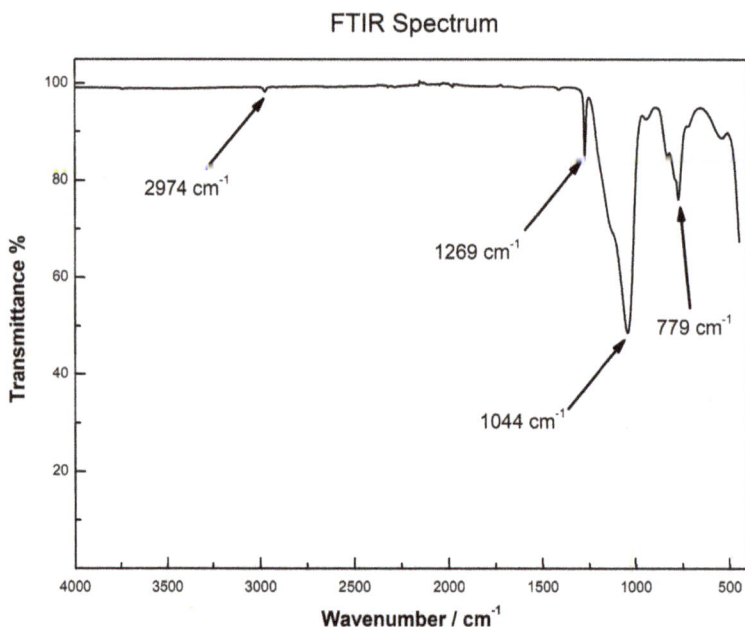

**Figure 2.** FTIR spectrum for the CW 20M stationary phase.

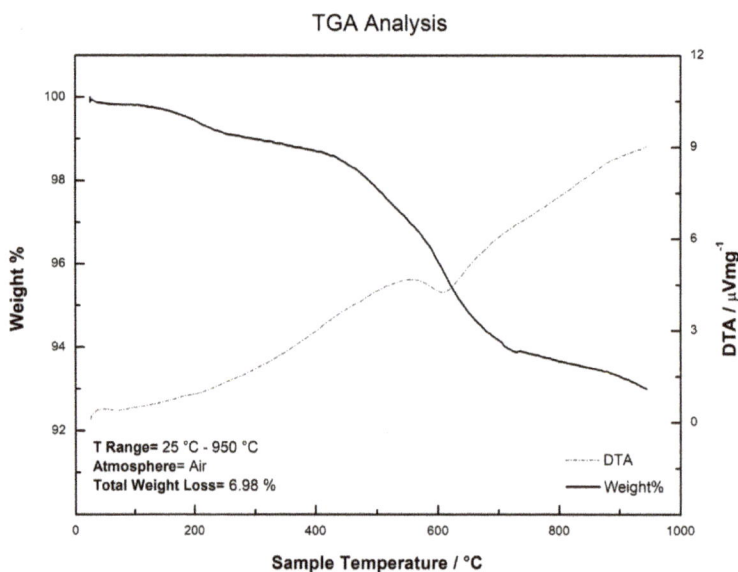

**Figure 3.** TGA analysis for the CW 20M stationary phase.

*2.4. Enrichment Factors Determination*

In Table 4 we reported the enrichment factors observed for the tested compounds. As highlighted within the SPE format, Oasis HLB shows the highest values (up to 607-folds), even if sol-gel CW 20M shows more reproducible values for all compounds, especially for cortisone. Contrary to the speculation made in Section 3.6, sol-gel CW 20M did not provide superior extraction performance compared to its commercial counterparts including C18 and HLB sorbents. The unexpected low performance can be attributed to the insufficient cleaning of the sol-gel CW 20M sorbent after the synthesis and conditioning (the TGA profile of sol-gel CW 20M corroborates this hypothesis). As such, the matrix still contains trapped solvent and unreacted reaction byproducts. Therefore, interaction between the target analytes and the sorbent occurred preferentially on the outer surface. Further experimentation is needed to evaluate actual extraction performance of the new sorbent after proper cleaning to ensure that all the trapped solvent and the unreacted reaction byproducts are exhaustively removed.

**Table 4.** Enrichment factors (%) observed at concentrations of 0.8, 2.5, and 8 µg/mL of water standard solutions. The enrichment factors were calculated as the percentage of peak area enhancement (after extraction) with respect to the area of reference standard solutions.

| Analyte | QC Concentration (µg/mL) | FPSE CW 20M | SPE Format | | | |
|---|---|---|---|---|---|---|
| | | | Avantor C18 | Sep-Pac Vac C18 | Oasis HLB | CW 20M |
| Ciprofloxacin | 0.8 | 28.4 | 740 | 304 | 67 | 321 |
| | 2.5 | 37.0 | 406 | 199 | 607 | 186 |
| | 8 | 13.6 | 200 | 188 | 367 | 113 |
| Sulfasalazine | 0.8 | 248 | 640 | 100 | 380 | 202 |
| | 2.5 | 164 | 325 | 206 | 408 | 217 |
| | 8 | 278 | 344 | 299 | 493 | 225 |
| Methyl-*p*-hydroxybenzoate | 0.8 | 190 | 204 | 78 | 314 | 59 |
| | 2.5 | 155 | 310 | 133 | 344 | 113 |
| | 8 | 121 | 205 | 216 | 252 | 76 |
| Cortisone | 0.8 | 136 | 80 | 123 | n.a. | 171 |
| | 2.5 | 88 | 160 | 195 | 167 | 199 |
| | 8 | 128 | 203 | 240 | 245 | 161 |

n.a. not available, the analyte is not detected during the analysis.

In Table 4 we also reported the enrichment factors observed for sol-gel CW 20M in the FPSE format. These values, in accordance with previously observed ones [6], were lower than the sol-gel CW 20M in SPE format, even if always comparable (or higher) in respect to other SPE commercially available devices.

It should be noted that, even if the enrichment factors calculated for both the SPE and the FPSE (Table 4) can vary a lot, and for the latter they were always lower than the SPE, the greatest advantage of the FPSE technique lies in the fact that it allows extraction of the analytes directly from the matrix of interest without having to resort to preliminary procedures of protein precipitation and/or cleaning up [4,6,7] to avoid the clogging phenomena.

## 3. Material and Methods

### 3.1. Chemicals, Solvents and Devices

Ciprofloxacin, methyl-*p*-hydroxybenzoate (methyl paraben, IS), sulfasalazine, and cortisone (>98% purity grade), sodium phosphate monobasic and sodium phosphate dibasic (>99% purity grade) and phosphoric acid were purchased from Sigma-Aldrich (Milan, Italy). Acetonitrile (ACN) and methanol (HPLC-grade), DMSO and HCl 37% (RPE-ISO for analysis) were purchased from Carlo Erba (Milan, Italy) and were used without further purification. The water (18.2 MΩ·cm at 25 °C) for HPLC analysis was generated by a Millipore Milli-Q Plus water treatment system (Millipore Bedford Corp, Bedford, MA, USA).

Solid phase extraction devices evaluated in this work were Oasis HLB (Waters, Milford, MA, USA), Sep-Pak Vac C18 (Waters, Milford, MA, USA), Avantor C18 loaded SPE (J.T. Baker, Avantor Performance Materials, LLC., Center Valley, PA, USA), and novel Carbowax 20M (sol-gel CW 20M, highly polar sorbent possessing poly(ethylene glycol), $H[OCH(CH_3)CH_2]_nOH$ as the building block).

All devices were in the format of 1 mL, 30 mg. GraphPad Prism v.4 (GraphPad Software Inc, San Diego, CA, USA) was used for the statistical analysis of experimental data.

### 3.2. Stock Solution and Quality Control Samples

The stock solutions of chemical standards were made at the concentration of 1 mg/mL in the appropriate solvent, referring to solubility tests: ciprofloxacin was solubilized in a mixture MeOH: HCl (0.1 M) 1:1 ($v/v$), sulfasalazine in DMSO, methyl-$p$-hydroxybenzoate and cortisone in MeOH. The combined working solutions (at the concentration of 10 and 50 µg/mL) were prepared by dilution of a mixed solution in MilliQ water. The resulting samples were used to evaluate the enrichment factors, breakthrough volume, retention volume, hold-up volume, retention factor, and theoretical plate number.

### 3.3. Apparatus and Chromatographic Conditions

Analyses were performed using an HPLC Thermo Fisher Scientific liquid chromatography system (Model: Spectra System P2000) coupled to a photodiode array detector (PDA) Model: Spectra System UV6000LP. Mobile phase was directly on-line degassed by using a Spectra System SCM1000 (Thermo Fisher Scientific, Waltham, MA, USA). Excalibur v.2.0 Software (Thermo Fisher Scientific, Waltham, MA, USA) was used to collect and analyze the data. The Luna C18 (250 × 4.6 mm, 5 µm particle size; Phenomenex, Torrance, CA, USA) packing column connected to a security guard column (4.0 × 3.0 mm, 5 µm particle size; Phenomenex, Torrance, CA, USA) were used to separate drugs and internal standard (IS). The column and security guard column were thermostated at 25 °C (± 1 °C) using a Jetstream2 Plus column oven during the analysis. Drugs and IS were detected at the maximum wavelengths of 283 nm (ciprofloxacin), 369 nm (sulfasalazine), 260 nm (methyl-$p$-hydroxybenzoate), 247 nm (cortisone), respectively, following the validated method reported in reference [6].

### 3.4. Preparation of Fabric Phase Sorptive Extraction (FPSE) Media

The preparation of the cellulose fabric substrate for sol-gel coating, the preparation of the sol solution for sol-gel coating, and the sol-gel immersion coating process have been described in detail elsewhere [15].

### 3.5. Synthesis of Sol-Gel CW 20M SPE Sorbent

The sol solution for sol-gel CW 20M SPE sorbent was prepared by mixing Carbowax 20M polymer, tetramethoxysilane (TMOS), methyl trimethoxysilane (MTMS), and isopropanol in a molar ratio 0.015:1:1:20, respectively, as reported by Kabir and Furton [17]. The mixture was first vortexed for 5 min and subsequently sonicated for 30 min to ensure that the resulting solution became homogeneous and free of any trapped bubbles. Afterwards, 0.1 M HCl was added to the mixture in a molar ratio of TMOS: 0.1 M HCl, 1:4. The mixture was mixed again for 3 min and the sol solution was left overnight thus that the hydrolysis of the sol-gel precursors, TMOS and MTMS, continued towards completion. Subsequently, 1 M NH$_4$OH was added in droplets to the sol solution under vigorous stirring until the pH of the solution became basic and the sol solution started turning into a gel spontaneously. Once the gelation completes, the sol-gel matrix was allowed to age for 48 h. The sol-gel matrix was then crushed and dried in a vacuum oven at 70 °C for 48 h. The dried sol-gel CW 20M SPE sorbent was then rinsed with 50:50 ($v/v$) methanol: Methylene chloride under sonication to remove unreacted sol solution ingredients and reaction byproducts. The sol-gel CW 20M sorbent was dried again at 70 °C for 24 h. Finally, the dried sol-gel CW 20M sorbent was pulverized in a ball mill into fine particles and particles between 40–52 µm diameter were collected in a mesh screen for using as the SPE sorbent.

### 3.6. Preparation of the SPE Media

The SPE cartridges used in this work were both commercial (Sep-Pak Vac C18, Oasis HLB, and Avantor C18) and in-house loaded with the sol-gel CW 20M (this latter stationary phase was synthetized as previously reported in paragraph 3.5). All devices were obtained in the format of 1 mL, 30 mg in order to compare the analytical performances directly. The principal SPE phase characteristics are reported in Table 5.

**Table 5.** Physical characteristics of SPE sorbent phases.

| Sorbent | Sorbent Mass (mg) | Surface Area ($m^2/g$) | Pore Diameter (Å) | Pore Volume ($cm^3/g$) | Particle Size (µm) |
|---------|-------------------|------------------------|-------------------|------------------------|--------------------|
| Oasis HLB | 30 | 800 | 80 | 1–3 | 30 |
| Sep-Pac Vac C18 | 30 | 325 | 125 | - | 55–105 |
| Avantor C18 | 30 * | 320–350 | 60 | - | 40 |
| Carbowax 20M | 30 | 990 | 71 | 1.8 | 40 |

* Commercially available in the size of 1 mL/100 mg sorbent phase; to compare cartridges with the same amount of sorbent phase, we have emptied and reconstituted them with 30 mg of their sorbent phase.

Compared to Brunauer–Emmett–Teller (BET) adsorption isotherm values (surface area, pore diameter, and pore volume) obtained for sol-gel CW 20M SPE sorbent with commercially available HLB and C18 sorbents, sol-gel CW 20M SPE sorbent demonstrated higher surface area and pore volume and is expected to provide superior extraction efficiency compared to their commercial counterparts.

### 3.7. SPE Sorbents Characterization Using FTIR and TGA

FTIR spectroscopy was conducted on a Perkin-Elmer spectrum TWO instrument, operating in Attenuated Total Reflectance (ATR) mode in the 4000 $cm^{-1}$–450 $cm^{-1}$ wavenumber range. Thermogravimetric analysis was carried out using a Netzsch Regulus 2500 thermobalance, in air atmosphere, in the 25–950 °C temperature range.

## 4. Conclusions

In these researches, it was highlighted that for the tested compounds nowadays there is not available a "universal" phase. In fact, for some analytes it was not possible to evaluate the breakthrough volume ($V_B$), retention volume ($V_R$), hold-up volume ($V_M$), retention factor ($k$), and theoretical plates number ($N$), because no quantitative data were obtained after the SPE steps. Consequently, it was not possible to perform the analysis by Boltzmann's function due to a "floating point error".

Comparing the breakthrough volume and enrichment factors data allowed us to evaluate the performance of both the FPSE and SPE techniques. The SPE technique showed the highest enrichment factors; consequently, this method is more suitable for samples with low analytes concentration. Indeed, breakthrough volumes and the number of theoretical plates data from the SPE method are higher for analytes concentration at 10 µg/mL, than the analytes concentration at 50 µg/mL. Generally, the deployment of silica sorbents (e.g., Avantor C18 and Sep-Pac-Vac C18) is not that advantageous relative to polymeric sorbents (e.g., Oasis HLB and sol-gel CW 20M) due to their instability in broader pH ranges. Additionally, silanol groups in silica sorbents bind irreversibly with target components, thus requiring more eluting solvents and a large consumption of time [18,19].

Among the different sorbent phases tested, Oasis HLB cartridges have the best performances, furthermore, against expectations, we found that even C18 cartridges showed better results than sol-gel CW 20M cartridge for that kind of analytes.

**Author Contributions:** Data curation, A.T., D.M., E.S., S.P., F.M. and P.B.; investigation, A.T., D.M., E.S., S.P., F.M. and P.B.; methodology, M.L., A.K., K.G.F., H.I.U., F.C. and V.F.S.; project administration, M.L., A.K., K.G.F., H.I.U., F.C. and V.F.S.; supervision, M.L., A.K., K.G.F., H.I.U., F.C. and V.F.S.

**Funding:** This research was funded by grant MIUR ex 60%, University of Chieti–Pescara "G. d'Annunzio", Chieti, Italy.

**Acknowledgments:** The authors want to thank Enrico Marcantoni and Serena Gabrielli at the Chemical Science Department—School of Science and Technology at the university of Camerino (Italy) for making available the FTIR and TGA instrumentations.

**Conflicts of Interest:** The authors declare no conflict of interest

# References

1. Samanidou, V.; Kaltzi, I.; Kabir, A.; Furton, K.G. Simplifying sample preparation using fabric phase sorptive extraction technique for the determination of benzodiazepines in blood serum by high-performance liquid chromatography. *Biomed. Chromatogr.* **2016**, *30*, 829–836. [CrossRef] [PubMed]
2. Wells, M.J.M. Handling large volume samples: applications of SPE to environmental matrices. In *Solid-Phase Extraction: Principles, Techniques, and Application*, 1st ed.; Simpson, N.K.J., Ed.; CRC Press: Boca Raton, FL, USA, 2000.
3. Kabir, A.; Furton, K.G. Fabric Phase Sorptive Extractors. United States Patents 9557252, 31 January 2017.
4. Locatelli, M.; Kabir, A.; Innosa, D.; Lopatriello, T.; Furton, K.G. A Fabric Phase Sorptive Extraction-High Performance Liquid Chromatography-Photo Diode Array Detection Method for the Determination of Twelve Azole Antimicrobial Drug Residues in Human Plasma and Urine. *J. Chromatogr. B* **2017**, *1040*, 192–198. [CrossRef] [PubMed]
5. Aznar, M.; Alfaro, P.; Nerin, C.; Kabir, A.; Furton, K.G. Fabric phase sorptive extraction: An innovative sample preparation approach applied to the analysis of specific migration from food packaging. *Anal. Chim. Acta* **2016**, *936*, 97–107. [CrossRef] [PubMed]
6. Kabir, A.; Furton, K.G.; Tinari, N.; Grossi, L.; Innosa, D.; Macerola, D.; Tartaglia, A.; Di Donato, V.; D'Ovidio, C.; Locatelli, M. Fabric Phase Sorptive Extraction-High Performance Liquid Chromatography-Photo Diode Array Detection Method for Simultaneous Monitoring of Three Inflammatory Bowel Disease Treatment Drugs in Whole Blood, Plasma and Urine. *J. Chromatogr. B* **2018**, *1084*, 53–63. [CrossRef] [PubMed]
7. Locatelli, M.; Tinari, N.; Grassadonia, A.; Tartaglia, A.; Macerola, D.; Piccolantonio, S.; Sperandio, E.; D'Ovidio, C.; Carradori, S.; Ulusoy, H.I.; et al. FPSE-HPLC-DAD method for the quantification of anticancer drugs in human whole blood, plasma and urine. *J. Chromatogr. B* **2018**, *1095*, 204–213. [CrossRef] [PubMed]
8. Bacalum, E.; Tanase, A.; David, V. Retention mechanisms applied in solid phase extraction for some polar compounds. *Analele Universitatii Bucaresti–Chimie* **2010**, *19*, 61–68.
9. Bielica-Daszkiewicz, K.; Voelkel, A. Theoretical and experimental methods of determination of the breakthrough volume of SPE sorbents. *Talanta* **2009**, *80*, 614–621. [CrossRef] [PubMed]
10. D'Angelo, V.; Tessari, F.; Bellagamba, G.; De Luca, E.; Cifelli, R.; Celia, C.; Primavera, R.; Di Francesco, M.; Paolino, D.; Di Marzio, L.; et al. MicroExtraction by Packed Sorbent and HPLC-PDA quantification of multiple anti-inflammatory drugs and fluoroquinolones in human plasma and urine. *J. Enz. Inhibit. Med. Chem.* **2016**, *31*, 110–116. [CrossRef] [PubMed]
11. Locatelli, M.; Ciavarella, M.T.; Paolino, D.; Celia, C.; Fiscarelli, E.; Ricciotti, G.; Pompilio, A.; Di Bonaventura, G.; Grande, R.; Zengin, G.; et al. Determination of Ciprofloxacin and Levofloxacin in Human Sputum Collected from Cystic Fibrosis Patients using Microextraction by Packed Sorbent-High Performance Liquid Chromatography Photo Diode Array Detector. *J. Chromatogr. A* **2015**, *1419*, 58–66. [CrossRef] [PubMed]
12. Locatelli, M.; Cifelli, R.; Di Legge, C.; Barbacane, R.C.; Costa, N.; Fresta, M.; Celia, C.; Capolupo, C.; Di Marzio, L. Simultaneous determination of Eperisone Hydrochloride and Paracetamol in mouse plasma by High Performance Liquid Chromatography-Photo Diode Array Detector. *J. Chromatogr. A* **2015**, *1388*, 79–86. [CrossRef] [PubMed]
13. Campestre, C.; Locatelli, M.; Guglielmi, P.; De Luca, E.; Bellagamba, G.; Menta, S.; Zengin, G.; Celia, C.; Di Marzio, L.; Carradori, S. Analysis of imidazoles and triazoles in biological samples after MicroExtraction by Packed Sorbent. *J. Enz. Inhibit. Med. Chem.* **2017**, *32*, 1053–1063. [CrossRef] [PubMed]

14. Bielica-Daszkiewicz, K.; Voelkel, A.; Rusińska-Roszak, D.; Zarzycki, P.K. Estimation of the breakthrough volume of selected steroids for C-18 solid-phase extraction sorbent using retention data from micro-thin layer chromatography. *J. Sep. Sci.* **2013**, *36*, 1104–1111. [CrossRef] [PubMed]

15. Kabir, A.; Mesa, R.; Jurmain, J.; Furton, K. Fabric Phase Sorptive Extraction Explained. *Separations* **2017**, *4*, 21. [CrossRef]

16. Smith, A.L. Infrared spectra-structure correlations for organosilicon compounds. *Spectrochim. Acta* **1960**, *16*, 87–105. [CrossRef]

17. Kabir, A.; Furton, K.G. Sol-Gel Polymeric Stationary Phases for High-Performance Liquid Chromatography and Solid Phase Extraction: Their Method of Making. U.S. Patent 9925518, 27 March 2018.

18. Raza, N.; Kim, K.-H.; Abdullah, M.; Raza, W.R.; Brown, J.C. Recent developments in analytical quantitation approaches for parabens in human-associated samples. *TrAC* **2018**, *98*, 161–173. [CrossRef]

19. Pavlović, D.M.; Babić, S.; Horvat, A.J.M.; Kaštelan-Macan, M. Sample preparation in analysis of pharmaceuticals. *TrAC* **2007**, *26*, 1062–1075.

**Sample Availability:** Samples of the compounds are available from the authors under request.

*molecules*

*Article*

# Rapid Monitoring of Organochlorine Pesticide Residues in Various Fruit Juices and Water Samples Using Fabric Phase Sorptive Extraction and Gas Chromatography-Mass Spectrometry

Ramandeep Kaur [1], Ripneel Kaur [1], Susheela Rani [1], Ashok Kumar Malik [1,*], Abuzar Kabir [2,*], Kenneth G. Furton [2] and Victoria F. Samanidou [3]

[1]  Department of Chemistry, Punjabi University, Patiala 147002, India; ramandeep.chem@gmail.com (R.K.); ripneel83chahal@gmail.com (R.K.); susheela.chemistry@gmail.com (S.R.)
[2]  Department of Chemistry and Biochemistry, International Forensic Research Institute, Florida International University, 11200 SW 8th St, Miami, FL 33199, USA; furtonk@fiu.edu
[3]  Laboratory of Analytical Chemistry, Department of Chemistry, Aristotle University of Thessaloniki, 54124 Thessaloniki, Greece; samanidu@chem.auth.gr
*   Correspondence: malik_chem2002@yahoo.co.uk (A.K.M.); akabir@fiu.edu (A.K.); Tel.: +91-175-3046598 (A.K.M.); +1-305-348-2396 (A.K.); Fax: +91-175-2283073 (A.K.M.); +1-305-348-4172 (A.K.)

Academic Editor: Victoria Samanidou
Received: 24 January 2019; Accepted: 6 March 2019; Published: 13 March 2019

**Abstract:** Fabric phase sorptive extraction, an innovative integration of solid phase extraction and solid phase microextraction principles, has been combined with gas chromatography-mass spectrometry for the rapid extraction and determination of nineteen organochlorine pesticides in various fruit juices and water samples. FPSE consolidates the advanced features of sol-gel derived extraction sorbents with the rich surface chemistry of cellulose fabric substrate, which could extract the target analytes directly from the complex sample matrices, substantially simplifying the sample preparation operation. Important FPSE parameters, including sorbent chemistry, extraction time, stirring speed, type and volume of back-extraction solvent, and back-extraction time have been optimized. Calibration curves were obtained in a concentration range of 0.1–500 ng/mL. Under optimum conditions, limits of detection were obtained in a range of 0.007–0.032 ng/mL with satisfactory precision (RSD < 6%). The relative recoveries obtained by spiking organochlorine pesticides in water and selected juice samples were in the range of 91.56–99.83%. The sorbent sol-gel poly(ethylene glycol)-poly(propylene glycol)-poly(ethylene glycol) was applied for the extraction and preconcentration of organochlorine pesticides in aqueous and fruit juice samples prior to analysis with gas chromatography-mass spectrometry. The results demonstrated that the present method is simple, rapid, and precise for the determination of organochlorine pesticides in aqueous samples.

**Keywords:** fabric phase sorptive extraction; gas chromatography-mass spectrometry; organochlorine pesticides; sample preparation

---

## 1. Introduction

Organochlorine pesticides (OCPs), a sub-class of persistent organic pollutants (POPs), have been mass-produced since the 1940s and widely applied in agriculture worldwide as important insecticides because of their cheaper price. They gained popularity due to their effectiveness in controlling mosquitoes, hence controlling malaria and typhoid fever [1,2]. OCPs are among nine of the initial "dirty dozen" persistent organic pollutants (POPs) identified by the Stockholm Convention on persistent organic pollutants (POPs) in 2001 [3]. As OCPs have low water solubility and high lipid solubility, they easily accumulate in the environment and living organisms. Additionally, these persistent environmental pollutants are very hazardous, and due to their volatility are susceptible to long-range atmospheric transport [4,5]. They are capable of biomagnifications through the food chain. Surface runoff from non-point sources, discharge of industrial wastewater, disposal of empty containers and equipment, discharge from surface application of pesticide [6] and careless washing lead in the addition of OCPs to the aquatic environment [7]. A study revealed that OCPs are responsible for toxic effects to aquatic organisms. Furthermore, they can accumulate in the ecosystem and potentially pose threat to biodiversity [8]. Despite the ban on their industrial production since the 1970s, OCPs are still found at trace levels in the environment due to their resilient physiochemical properties. OCPs have been found to cause several carcinogenic and non-carcinogenic disorders in humans [9,10]. Increasing concern regarding health safety issues has emphasized the need for detection of pesticides at trace levels in the drinking water and food as these have leached the soil and entered the food chain [11]. Therefore, detection of OCP residues is of utmost importance in estimating potential health risks, performing ecotoxicological risk assessments, and enforcing regulations [12,13]. The development of simple, fast, reliable and environmentally friendly methods that enable the determination of OCPS at trace levels in aqueous samples is of great concern nowadays [14,15].

For the trace analysis of target analytes in various matrices, the sample preparation is of paramount importance to analytical efficiency and analyte recovery. Various extraction techniques based on solid sorbents have been extensively used for the determination of OCPs from aqueous samples including solid-phase extraction (SPE) [5,16], solid-phase microextraction (SPME) [17], single drop microextraction (SDME) [18], magnetic solid-phase extraction (MSPE) [19], micro solid-phase extraction (μ-SPE) [20,21] and stir-bar sorptive extraction (SBSE) [22,23]. The extraction of analytes relies on the type of sorbent material being used. Moreover, selectivity of the extraction method depends upon the nature of the sorbent, which determines their affinity towards the target compounds. In this regard, sol-gel technology has achieved considerable success in analytical sciences, with potential applications in the adsorption and separation of various analytes which result from their distinct features, such as unique selectivity, enhanced extraction sensitivity and higher thermal, mechanical and solvent stability [24]. FPSE, introduced in 2014 by Kabir and Furton, represents an important development for the extraction of several organic pollutants at trace and ultra trace levels. It combines the sampling, isolation and enrichment of the analytes in a single step, making it a quick sample preparation process [25,26]. FPSE consists of a small cellulose or polyester fabric coated with a thin layer of a suitable sorbent phase by sol-gel coating technology. There is a large number of coatings available for FPSE, and the extraction performance is dependent on the choice of an appropriate sorbent [27,28]. FPSE is an equilibrium technique based on partitioning of the analytes between the matrix and an extraction phase. The FPSE media is inserted into the sample solution, which is agitated for a certain time, and the adsorbent with the adsorbed analyte is then separated from the solution [29]. The analytes are consequently eluted and analyzed. FPSE possesses some exceptional and unprecedented properties such as ultra-high specific surface area, increased surface activities, flexible functionalization and tunable composition, which make them suitable for versatile and efficient sample preparation [30–32]. Additionally, FPSE media are very stable in a wider range of harsh organic solvent due to stronger chemical bonding between sol-gel sorbent and fabric media [33]. One of the most crucial factors for successful application of FPSE is direct extraction of analyte from real sample matrices without any specific requirement, such as filtration, centrifugation, solvent evaporation and sample reconstitution [34–36].

The present study involved an exploratory study on the feasibility of using FPSE media for the extraction and determination of persistent organic pollutant in various real sample matrices. This paper demonstrates a simple, rapid and efficient FPSE method coupled with GC-MS that we have developed to analyze the amount of organochlorine pesticides in water and juice samples obtained from market.

The optimization of the analytical process, including the influence of several experimental parameters, selectivity, and interaction mechanisms, is fully discussed.

## 2. Results and Discussion

### 2.1. Optimization of Fabric Phase Extraction

To evaluate the capability of FPSE media, the extraction of a mixture of organochlorine containing 19 OCPs as model compounds from water and juice samples were investigated. Several FPSE parameters, including the extraction time, stirring speed, desorption solvent and its volume, desorption time, and ionic strength, were investigated to achieve the best extraction efficiency of the chosen FPSE media.

#### 2.1.1. Selection of Fabric Phase Sorptive Extraction Sorbent Chemistry

One of the most important tasks in fabric phase sorptive extraction method development is to select the appropriate sorbent chemistry that would offer the highest selectivity and extraction efficiency towards the target analytes. Failure to select the appropriate sorbent chemistry may substantially limit the overall sensitivity of the analytical method. Unlike other microextraction techniques, which have a limited number of sorbents to choose from, fabric phase sorptive extraction offers a large number of sol-gel-based high-efficiency sorbents that can potentially be used for a given application. It would be an utterly tedious job if one had to test all the available sorbents, as in the case of most sample preparation techniques, to find the most suitable one. To simplify the sorbent selection process, fabric phase sorptive extraction has developed an absolute recovery prediction calculator using the logKow values of the analytes for each of the FPSE sorbent media. Using the logKow value of the analyte, one can predict about the tentative absolute recovery of the particular analyte. For multi-residual analysis, once the recovery value for each of the analyte is calculated, the FPSE sorbent, which provides the highest recovery values for most of the analytes, can be selected or two/three FPSE sorbent media, can be short-listed for method development and the best one can be selected from the experimental data. Taking the medium and low polarity of the organochlorine pesticides into consideration, two medium polar FPSE coatings, sol-gel poly(ethylene glycol)-poly(propylene glycol)-poly(ethylene glycol) (sol-gel PEG-PPG-PEG) and sol-gel poly(caprolactone)-poly(dimethyl siloxane)-poly(caprolactone) (sol-gel PCAP-PDMS-PCAP) were selected as the potential sorbent candidates. The schematic representation of sol-gel PEG-PPG-PEG and sol-gel PCAP-PDMS-PCAP coated FPSE media are presented in Figure 1.

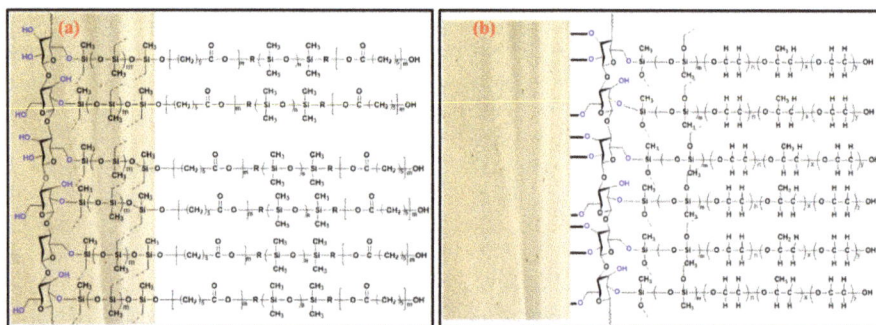

**Figure 1.** Schematic representation of (**a**) sol-gel PCAP-PDMS-PCAP; (**b**) sol-gel PEG-PPG-PEG coated FPSE media.

The absolute recovery equations for both the sorbents are given below:

Absolute Recovery, % for sol-gel PEG-PPG-PEG: $-3.68 + 23.07 \log Kow - 1.72 (\log Kow - 2.74)^2$

Absolute Recovery, % for sol-gel PCAP-PDMS-PCAP: $-3.36 + 22.34 \log Kow - 2.12 (\log Kow - 2.74)^2$

Since the model is valid between logKow 0.3–5.07, the predicted absolute recovery values have been calculated for only 7 OCPs whose logKow values are within this range. The predicted absolute recovery values are presented in Table 1.

**Table 1.** Predicted absolute recovery values (%) for selected OCPs on two FPSE media.

| FPSE Sorbent | Predicted Absolute Recovery Values (%) | | | | | | |
|---|---|---|---|---|---|---|---|
| | β-Endosulfan | Endosulfan Sulfate | γ-BHC | β-BHC | α-BHC | δ-BHC | Endrine Aldehyde |
| Sol-gel PEG-PPG-PEG | 74.49 | 75.29 | 76.48 | 77.65 | 78.37 | 84.44 | **95.71** |
| Sol-gel PCAP-PDMS-PCAP | **75.86** | **76.61** | **77.70** | **78.79** | **79.33** | **84.97** | 94.88 |

The highest value obtained for each analyte is presented in bold. As seen in Table 1, the sol-gel PCAP-PDMS-PCAP FPSE sorbent projected higher recovery for almost all of the analytes. Based on this prediction, both FPSE sorbents were subjected to the extraction of OCPs.

Both the FPSE media were used for the extraction of the target OCPs from 10 mL spiked aqueous solution at 5 ng/mL, extraction time, 20 min; stirring speed at 900 rpm, back-extraction in acetone for 15 min. The results are presented in Figure 2a. Although the absolute recovery model predicted that sol-gel PCAP-PDMS-PCAP would perform better than sol-gel PEG-PPG-PEG, in reality, the extraction performance for all OCPs were superior in sol-gel PEG-PPG-PEG FPSE media. The root cause for this discrepancy needs further investigation and perhaps refinement of the absolute recovery calculator. Nonetheless, the models can definitely help in shortlisting suitable FPSE sorbents for selecting the best from real experimentations. Based on the experimental results, sol-gel PEG-PPG-PEG coated FPSE media was selected as the optimum sorbent for OCPs and was used in all the subsequent method development and validation experiments.

### 2.1.2. Effect of Stirring Speed

Sample agitation is a critical parameter in the extraction process according to mass transfer theory. Sample agitation increases the movement of analytes to fabric surface with reduction in thickness of boundary layer and shortened thermodynamic equilibrium time. Therefore, the effect of stirring speed on the extraction efficiency of the analytes was investigated within the interval of 300–1200 rpm. The extraction efficiency of the analytes increased with increase in the stirring speed as demonstrated by the results (Figure 2b). The highest extraction efficiency was achieved at a stirring speed of 1200 rpm. Therefore, 1200 rpm stirring speed was selected for the subsequent experiments.

### 2.1.3. Effect of Extraction Time

The extraction time is a critical parameter in the FPSE procedure as it greatly influences the partition of the target analytes between the sample solution and the sorbent. FPSE is an equilibrium-based technique, and the mass transfer of analytes from the sample solution to the sorbent is directly influenced by the extraction time. To achieve the highest extraction recovery, the effect of time on the extraction efficiency was examined in the range of 5–50 min. The adsorbed amounts of OCPs generally increased with extraction time up to 30 min with no obvious change occurring thereafter as depicted in Figure 2c. Thus, the extraction reached the equilibrium at 30 min. Consequently, an extraction time of 30 min was selected for further analysis.

### 2.1.4. Effect of Matrix pH

OCPs are persistent organic pollutants that are present in the neutral state within the entire pH range in an aqueous solution. Hence, the pH of the sample solution was not expected to have a significant impact on the extraction efficiency. Nonetheless, the effect of pH on recovery was conducted by varying pH values ranged from 2 to 10 adjusting with HCl or NaOH ($n = 3$). Indeed, no significant difference in terms of extraction efficiency was observed for OCPs. Hence, subsequent experiments were conducted with measured pH values in the range of 6.0–7.0 without any pH modification in the sample solution.

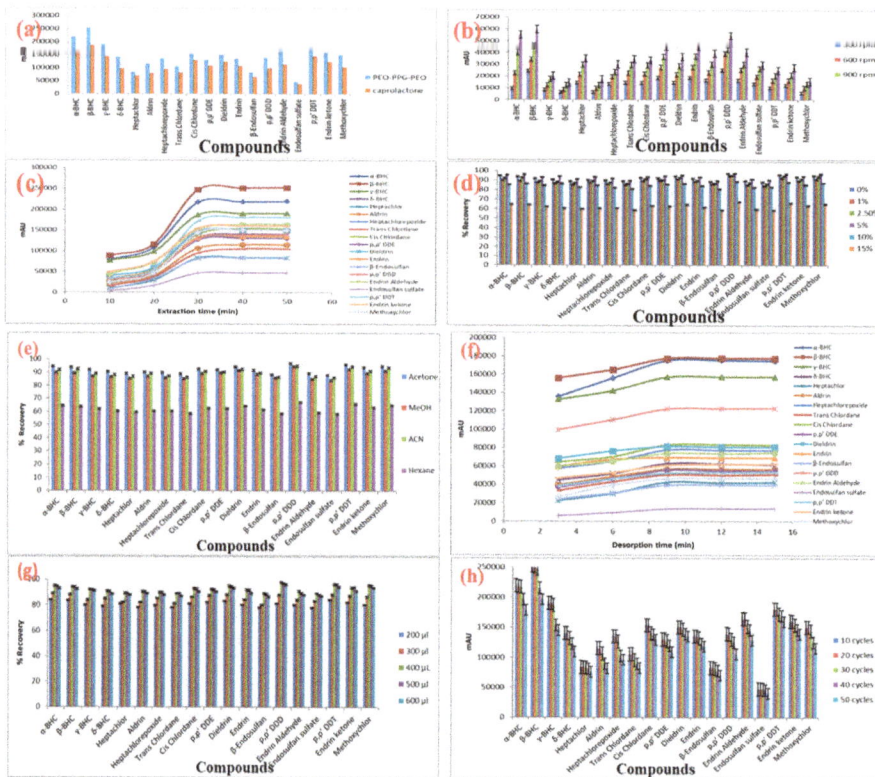

**Figure 2.** Effect of (**a**) sorbent type; (**b**) agitation speed; (**c**) extraction time; (**d**) salt addition; (**e**) desorption solvent; (**f**) desorption time; (**g**) desorption solvent volume; (**h**) repeated use of FPSE media on the microextraction efficiency of the analytes (sample volume: 10 mL; concentration of the analytes 5 ng/mL).

## 2.1.5. Effect of Salt Addition

The addition of salt can enhance the ionic strength of the sample solution and diminish the solubility of analytes in the sample solution by salting out effect. Thus, the partitioning of the analytes from the sample solution to the adsorbent is revamped. In this study, the effect of ion strength on the extraction efficiency of the analytes was investigated by adding different concentrations 0, 1, 2.5, 5, 10 and 15%, (*w/v*) of NaCl into the standard working solution with three replicates at each point. The results revealed that the addition of NaCl up to 5.0% (*w/v*) increases the extraction efficiency due to the salting out effect and then decreases. Further increase in NaCl concentration can increase density and viscosity of the aqueous solution that diminishes the salting-out effect and results in low mass transfer and extraction efficiency (Figure 2d). Thus, 5.0% (*w/v*) was chosen as the suitable amount of NaCl for the subsequent experiment.

### 2.1.6. Desorption Conditions

The selection of an appropriate desorption solvent is necessary for retrieving analytes entrapped in the FPSE membrane to achieve higher extraction recovery. The process of desorption was carried out using different organic solvents, including acetonitrile, methanol, *n*-hexane and acetone. Acetone was proven to yield the highest recovery of analytes by desorbing the membrane immersed in 500 µL of solvent as shown in Figure 2e. Therefore, acetone was selected as the eluent for subsequent experiment. Thereafter, the minimum time required for the complete desorption of the analytes from the sorbent was also investigated in the range of 3–15 min. The extraction efficiency was the highest with adsorption time of 9 min. There was no improvement in the amount of OCPs present in the enriched solvent that was observed when desorption time was prolonged further to 15 or 20 min. Consequently, 9 min was selected as the optimal desorption time for subsequent experiments (Figure 2f).

To evaluate the effect of the solvent volume for complete desorption of the analyte on the extraction recovery, the volume of six different solvent volumes of acetone containing 10 ng/mL OCPs were investigated using the proposed approach. As shown in Figure 2g, the analytical signals for all the OCPs increased and reached their optimum levels when the solvent volume increased from 100 to 400 µL. The extraction sensitivity reduces when the solvent volume was further increased to 500 and 600 µL. Based on the results obtained, a desorption time of 9 min was selected for the subsequent analysis with desorption solvent volume at 400 µL.

### 2.1.7. Stability and Reusability of Sol-Gel FPSE Media

The stability and reusability of sorbent are the crucial features for better performance of the sorbent. For this purpose, the extraction efficiency, reusability and stability of the sorbents were assessed through thirty consecutive cycles of adsorption/desorption for OCPs extraction.

The stability of the sol-gel PEG-PPG-PEG coated FPSE media was evaluated based on reproducibility of extraction efficiency with different batches of sorbent. The reproducibility of the extraction was examined using five different batches of sol-gel PEG-PPG-PEG sorbent-coated FPSE media with water samples spiked with 5 ng/mL of each OCP. RSD values lower than 6% were obtained, indicating a good reproducibility in the sol-gel sorbent coating process. The regeneration of the FPSE media (for multiple use) was examined with a random batch. After the completion of each extraction-desorption cycle, the used FPSE media was regenerated by rinsing it with 0.5 mL of water and acetone to remove any residual or other substances. The regenerated sorbent was then inserted into the glass vial containing water spiked with OCPs. There is no significant change in the peak areas for each analyte extracted by the FPSE media even after 30 times of recycling as demonstrated in Figure 2h. The results proved that FPSE media are durable and stable with excellent reusability due to the strong bonding between cellulose and sol-gel sorbent coating.

## 2.2. Method Validation

### 2.2.1. Limit of Detection and Quantification

The proposed method was validated by figures of merit under the optimized experimental conditions (30 min of extraction time, pH 6–7, 9 min desorption time, 5% (*w/v*) of NaCl and 400 µL of acetone as desorption solvent agitated at 1200 rpm). The calibration curves were obtained by eight different concentrations of the OCPs' standard solutions. Good linearity was observed over the wide concentration ranges for the nineteen OPPs with satisfactory determination coefficients (r2). Correlation coefficients (r2) ranging from 0.9929 to 0.9986 were obtained for all the analytes. The LOD and LOQ values were obtained based on a signal-to-noise ratio of 3 and 10, respectively. The instrumental limits of detection (LODs) (S/N = 3) and quantification (LOQs) (S/N = 10) are listed in Table 2. The LODs and LOQs are in the range of 0.007–0.032 ng/mL and 0.023–0.069 ng/mL for all analytes, respectively.

**Table 2.** The performance characteristics of the proposed FPSE/GC-MS analytical method.

| Organochlorine Pesticides | Linear Range (ng/mL) | Coefficient of Determination, $r^2$ | LOD (ng/mL) | LOQ (ng/mL) | RSD % | |
|---|---|---|---|---|---|---|
| | | | | | Intra-Day | Inter-Day |
| α-Benzenehexachloride (α-BHC) | 0.1–500 | 0.9977 | 0.013 | 0.042 | 4.2 | 5.2 |
| β-Benzenehexachloride (β-BHC) | 0.1–500 | 0.9968 | 0.008 | 0.026 | 3.3 | 4.1 |
| γ-Benzenehexachloride (γ-BHC) | 0.1–500 | 0.9944 | 0.021 | 0.069 | 4.6 | 5.0 |
| δ-Benzenehexachloride (δ-BHC) | 0.1–500 | 0.9931 | 0.032 | 0.105 | 3.5 | 4.7 |
| Heptachlor | 0.1–500 | 0.9951 | 0.014 | 0.046 | 4.3 | 5.4 |
| Aldrin | 0.1–500 | 0.9929 | 0.026 | 0.086 | 2.3 | 3.8 |
| Heptachlorepoxide | 0.1–500 | 0.9965 | 0.015 | 0.049 | 3.3 | 4.5 |
| Trans Chlordane | 0.1–500 | 0.9952 | 0.013 | 0.042 | 3.1 | 3.9 |
| Cis Chlordane | 0.1–500 | 0.9954 | 0.014 | 0.046 | 2.6 | 3.3 |
| p,p Dichlorodiphenyldichloroeth ylene (p,p' DDE) | 0.1–500 | 0.9984 | 0.011 | 0.0363 | 2.7 | 3.7 |
| Dieldrin | 0.1–500 | 0.9963 | 0.013 | 0.042 | 3.5 | 5.3 |
| Endrin | 0.1–500 | 0.9938 | 0.012 | 0.039 | 3.4 | 5.6 |
| β-Endosulfan | 0.1–500 | 0.9960 | 0.016 | 0.053 | 2.8 | 3.7 |
| p,p Dichlorodiphenyldichloroeth ane (p,p' DDD) | 0.1–500 | 0.9969 | 0.007 | 0.023 | 3.7 | 4.2 |
| Endrin Aldehyde | 0.1–500 | 0.9977 | 0.012 | 0.039 | 3.1 | 3.9 |
| Endosulfan sulfate | 0.1–500 | 0.9932 | 0.015 | 0.049 | 4.6 | 5.5 |
| p,p Dichlorodiphenyltrichloroeth ane (p,p' DDT) | 0.1–500 | 0.9984 | 0.021 | 0.069 | 2.5 | 3.9 |
| Endrin ketone | 0.1–500 | 0.9986 | 0.018 | 0.059 | 4.1 | 5.7 |
| Methoxychlor | 0.1–500 | 0.9971 | 0.027 | 0.089 | 4.4 | 5.6 |

### 2.2.2. Precision and Accuracy

The precision, in terms of the relative standard deviations was also evaluated by performing five extraction replicates for each of the spiked OCPs at three different concentrations. While the intraday precision was obtained by determining the analytes five times in the same day, and the inter-day precision was obtained by performing the same procedure in five consecutive days. The recoveries and RSD values are shown in Table 2. The intra-day and inter-day RSD values were 2.3–4.6 and 3.3–5.7, respectively, at all concentrations. It is evident from the low RSD values that the developed FPSE/GCMS method is reliable and reproducible for the analysis of OCPs.

### 2.2.3. Application to Real Samples

To examine the applicability of the new FPSE/GC-MS method in real samples, the proposed sol-gel FPSE media coated with sol-gel PEG-PPG-PEG were used to extract and enrich OCPs from water and juice samples. Under the optimized experimental conditions, FPSE media coupled with GC-MS method was validated for the enrichment and determination of OCPs in real samples. The OCP levels in environmental water and drink samples, including tap water, ground water, municipal water, apple juice, pomegranate juice and litchi juice, were analyzed using the FPSE/GC-MS method developed in this study, and the results are summarized in Table 3. Four levels of OCP concentrations (0.1, 1, 10 and 100 ng/mL) were spiked in the actual samples to further test the applicability of the developed method. The spiking recoveries of the target OCPs in the four types of samples are listed in Table 3. The recoveries for the spiked samples ranged from 91.56% to 99.83%. The extracted ion chromatograms of the OCPs acquired from tap water, ground water, municipal water, apple juice, pomegranate juice and litchi juice samples through the developed FPSE/GC-MS method are shown in Figure 3.

**Table 3.** Analytical data obtained from FPSE/GC-MS analysis for the determination of OCPs in water and fruit juice samples.

| OCP | Amount Added ng/mL | Tap Water | | | Ground Water | | | Municipal Water | | | Apple Juice | | | Litchi Juice | | | Pomegranate Juice | | |
|---|---|---|---|---|---|---|---|---|---|---|---|---|---|---|---|---|---|---|---|
| | | Extraction Yield | Intraday RSD (%) | Interday RSD (%) | Extraction Yield | Intraday RSD (%) | Interday RSD (%) | Extraction Yield | Intraday RSD (%) | Interday RSD (%) | Extraction Yield | Intraday RSD (%) | Interday RSD (%) | Extraction Yield | Intraday RSD (%) | Interday RSD (%) | Extraction Yield | Intraday RSD (%) | Interday RSD (%) |
| α-Benzenehexachloride (α-BHC) | 0.1 | 97.4 | 4.5 | 5.1 | 96.8 | 4.7 | 4.9 | 96.6 | 3.5 | 4.8 | 95.7 | 4.3 | 5.4 | 95.6 | 3.6 | 4.4 | 95.5 | 4.4 | 5.6 |
| | 1 | 98.7 | 4.1 | 4.9 | 97.6 | 4.1 | 4.8 | 97.5 | 3.4 | 4.6 | 96.4 | 4.2 | 5.1 | 96.2 | 3.5 | 4.7 | 96.3 | 4.2 | 5.2 |
| | 10 | 98.9 | 3.6 | 4.3 | 97.8 | 4.2 | 5.2 | 97.3 | 3.2 | 4.2 | 96.7 | 3.7 | 4.8 | 97.1 | 3.4 | 4.3 | 96.3 | 3.3 | 4.9 |
| | 100 | 99.3 | 3.1 | 3.9 | 98.7 | 3.6 | 4.1 | 98.4 | 2.5 | 3.1 | 97.7 | 3.4 | 4.6 | 97.4 | 2.3 | 3.5 | 97.6 | 3.2 | 4.5 |
| β-Benzenehexachloride (β-BHC) | 0.1 | 96.4 | 4.2 | 5.4 | 97.4 | 4.1 | 4.9 | 95.6 | 4.7 | 5.7 | 94.4 | 3.7 | 5.6 | 96.9 | 4.9 | 5.5 | 95.5 | 4.2 | 5.3 |
| | 1 | 97.7 | 3.5 | 5.1 | 98.7 | 3.8 | 4.7 | 95.5 | 4.1 | 3.6 | 94.9 | 3.6 | 4.9 | 97.2 | 4.6 | 5.1 | 96.4 | 4.1 | 5.1 |
| | 10 | 98.4 | 3.1 | 4.7 | 98.6 | 3.6 | 4.2 | 96.3 | 4.2 | 3.6 | 95.3 | 3 | 4.5 | 97.5 | 4 | 4.9 | 96.8 | 3.3 | 4.7 |
| | 100 | 99.4 | 2.9 | 3.8 | 99.5 | 2.5 | 3.8 | 98.4 | 3.6 | 4.7 | 96.7 | 2.8 | 3.8 | 98.4 | 3.9 | 4.6 | 97.1 | 3.1 | 4.3 |
| γ-Benzenehexachloride (γ-BHC) | 0.1 | 97 | 3.9 | 5.1 | 97.4 | 3.6 | 4.7 | 96.6 | 4.2 | 5.3 | 96.6 | 4.6 | 5.5 | 96.2 | 4.9 | 5.7 | 95.6 | 4.7 | 5.7 |
| | 1 | 97.8 | 3.5 | 4.9 | 98.7 | 3.3 | 4.1 | 97.8 | 3.9 | 4.2 | 96.5 | 3.9 | 4.3 | 97.3 | 4.7 | 5.1 | 96.9 | 4.2 | 5.2 |
| | 10 | 98.9 | 3.4 | 4.3 | 98.9 | 3.1 | 4.2 | 97.4 | 3.4 | 4.8 | 96.9 | 3.4 | 3.9 | 97.5 | 3.9 | 4.6 | 97.3 | 3.9 | 4.9 |
| | 100 | 99.5 | 2.9 | 3.9 | 99.8 | 2.9 | 3.4 | 98.5 | 2.7 | 3.6 | 97.5 | 3 | 4.3 | 98.6 | 3.5 | 4.2 | 97.6 | 3 | 4.3 |
| δ-Benzenehexachloride (δ-BHC) | 0.1 | 97.6 | 3.7 | 5.3 | 97.9 | 3.8 | 5.1 | 97.9 | 4.8 | 5.8 | 95.6 | 3.7 | 5.1 | 93.6 | 4.8 | 5.6 | 91.5 | 4.4 | 5.1 |
| | 1 | 98 | 3.4 | 4.9 | 98 | 3.4 | 4.6 | 97.6 | 3.7 | 5.1 | 95.9 | 3.2 | 4.8 | 94.2 | 3.8 | 4 | 92.3 | 3.9 | 4.8 |
| | 10 | 98.7 | 3.3 | 4.2 | 98.6 | 2.8 | 3.9 | 98.5 | 3.1 | 4.8 | 96.4 | 2.9 | 3.8 | 97.1 | 3.2 | 3.3 | 94.3 | 3.5 | 4.1 |
| | 100 | 99.5 | 3 | 3.7 | 99.8 | 2.4 | 3.2 | 98.4 | 2.9 | 3.7 | 96.9 | 2.1 | 3.5 | 98.4 | 2.7 | 3.5 | 96.6 | 2.7 | 3.5 |
| Heptachlor | 0.1 | 97.2 | 3.8 | 5.1 | 97.6 | 4.7 | 5.7 | 97.5 | 4.9 | 5.9 | 96.7 | 3.6 | 5 | 97.9 | 4.6 | 5.4 | 97.6 | 4.4 | 5.3 |
| | 1 | 97.6 | 3.4 | 4.9 | 97.5 | 4.1 | 5 | 97.8 | 4.6 | 5.1 | 97.4 | 3.1 | 4.7 | 97 | 3.6 | 4.4 | 97.9 | 3.5 | 4.3 |
| | 10 | 98.1 | 3.3 | 4.3 | 98.3 | 4.2 | 4.9 | 98.4 | 4.2 | 4.9 | 97.7 | 2.7 | 3.9 | 98.1 | 3.4 | 4 | 98.1 | 3.3 | 3.9 |
| | 100 | 99.1 | 2.9 | 3.9 | 99.1 | 3.6 | 3.9 | 98.5 | 3.6 | 4.3 | 98.7 | 2.2 | 3.4 | 98.9 | 2.9 | 3.2 | 98.6 | 2.8 | 3.3 |
| Aldrin | 0.1 | 97.6 | 4.6 | 5.7 | 97.4 | 4.9 | 5.6 | 97.6 | 4.8 | 5.7 | 97.7 | 4.2 | 4.9 | 96.9 | 4.8 | 5.6 | 98.1 | 4.6 | 5.4 |
| | 1 | 97.6 | 4.2 | 5.2 | 97.2 | 4.2 | 4.9 | 97.9 | 4.3 | 5 | 98.4 | 3.8 | 4.5 | 97.1 | 4.2 | 4.9 | 98.3 | 3.8 | 4.1 |
| | 10 | 98.4 | 3.9 | 4.3 | 98.8 | 3.9 | 4.2 | 98.3 | 4 | 4.8 | 98.6 | 3.3 | 3.8 | 97.9 | 3.9 | 4.6 | 99 | 3.5 | 3.7 |
| | 100 | 99.3 | 3.2 | 3.9 | 98.5 | 2.8 | 3.5 | 99.4 | 3.1 | 4.2 | 99.4 | 2.9 | 3.2 | 98.1 | 3.3 | 3.9 | 99.6 | 2.9 | 3.5 |
| Heptachloroepoxide | 0.1 | 97.5 | 4.3 | 5 | 96.5 | 3.8 | 4.6 | 97.5 | 4.7 | 5.6 | 97 | 4.1 | 4.8 | 96.9 | 4.7 | 5.5 | 97.5 | 4.5 | 5.3 |
| | 1 | 98.8 | 3.6 | 4.8 | 97.7 | 3.3 | 4 | 97.6 | 4.2 | 5.1 | 97.2 | 3.7 | 4.4 | 97.1 | 4.1 | 5 | 97.9 | 3.6 | 4.2 |
| | 10 | 98.9 | 3.2 | 4.2 | 98.8 | 3 | 4.1 | 98.6 | 4.1 | 4.7 | 98.1 | 3.4 | 3.7 | 97.6 | 3.6 | 4.5 | 98.3 | 3 | 3.6 |
| | 100 | 99.1 | 3 | 3.8 | 98.6 | 2.3 | 3.5 | 99.3 | 2.9 | 4.1 | 98.8 | 2.5 | 3.1 | 98.6 | 3.1 | 3.9 | 98.7 | 2.8 | 3.4 |
| Trans Chlordane | 0.1 | 97.4 | 3.8 | 4.9 | 97.6 | 3.9 | 4.2 | 97.6 | 4.6 | 5.4 | 97 | 4.5 | 5.8 | 96.6 | 4.6 | 5.8 | 97.5 | 4.4 | 5.3 |
| | 1 | 97.9 | 3.7 | 4.5 | 98 | 3.6 | 3.7 | 97.8 | 4 | 4.9 | 97.9 | 3.9 | 4.4 | 97.1 | 4 | 5.1 | 98.5 | 3.7 | 4.3 |
| | 10 | 98.6 | 3.6 | 4.2 | 98.6 | 2.7 | 3.7 | 98.8 | 4 | 4.8 | 98.2 | 3.6 | 3.9 | 97.4 | 3.7 | 4.5 | 98.7 | 3.1 | 3.6 |
| | 100 | 99.2 | 2.3 | 3.8 | 99.8 | 2.4 | 2.6 | 99.4 | 3.7 | 4.5 | 99.6 | 2.5 | 3.1 | 98.3 | 3.2 | 3.8 | 99.1 | 2.9 | 3.4 |
| Cis Chlordane | 0.1 | 97.4 | 3.7 | 5.1 | 97.2 | 4.7 | 4.7 | 96.7 | 4.5 | 5.3 | 96.4 | 4.6 | 5.5 | 96.2 | 4.7 | 5.7 | 96.6 | 4.3 | 5.2 |
| | 1 | 98.9 | 3.4 | 4.9 | 98.1 | 4.1 | 4.1 | 97.8 | 4 | 4.9 | 96.4 | 4.3 | 4.9 | 97.1 | 4.1 | 5 | 97.4 | 3.6 | 4.3 |
| | 10 | 98.7 | 3.3 | 4.3 | 98.6 | 4.2 | 4.3 | 98.5 | 3.9 | 4.4 | 97.2 | 3.7 | 4.8 | 97.8 | 3.8 | 4.4 | 98.6 | 3.1 | 3.9 |
| | 100 | 99.3 | 3 | 3.9 | 99.3 | 3.6 | 3.9 | 99.4 | 2.9 | 3.5 | 98.8 | 2.6 | 3.2 | 98.2 | 3.3 | 3.1 | 98.2 | 2.8 | 3.5 |
| p,p Dichlorodiphenyldichloroethylene (p,p'DDE) | 0.1 | 96.5 | 3.6 | 5 | 97.2 | 4.8 | 5.2 | 96.5 | 4.6 | 5.6 | 96.7 | 4.9 | 5.6 | 96.8 | 4.6 | 5.5 | 96.5 | 4.5 | 5.1 |
| | 1 | 96.9 | 3.5 | 4.8 | 98.1 | 4.9 | 4.3 | 98 | 4.2 | 4.9 | 96.5 | 4.4 | 4.9 | 97 | 4 | 5 | 97.4 | 3.5 | 4.5 |
| | 10 | 97.7 | 3.2 | 4.2 | 98.6 | 4.7 | 4.2 | 98.9 | 3.8 | 4.8 | 97.1 | 3.9 | 4.7 | 97.5 | 3.9 | 4.3 | 98.5 | 3.1 | 3.9 |
| | 100 | 98.3 | 3 | 3.8 | 99.3 | 3.9 | 3.1 | 99 | 2.7 | 3.5 | 97.8 | 2.5 | 3.3 | 98 | 3.4 | 3.2 | 98.8 | 2.7 | 3.6 |
| Dieldrin | 0.1 | 94.8 | 4.9 | 5.8 | 95.7 | 4.7 | 5.6 | 94.5 | 4.5 | 5.5 | 95.7 | 4.4 | 5.5 | 95.6 | 4.1 | 5.3 | 94 | 4.3 | 5.2 |
| | 1 | 95.2 | 3.7 | 4.6 | 96.1 | 4.2 | 5.1 | 95 | 3.8 | 4.9 | 96.6 | 4.1 | 5.1 | 96.4 | 3.8 | 4.7 | 94.9 | 3.9 | 4.8 |
| | 10 | 95.6 | 3.3 | 4.2 | 97.2 | 2.9 | 4.2 | 95.8 | 2.9 | 3.7 | 97.1 | 3.9 | 4.6 | 97.6 | 3.6 | 4.5 | 95 | 3.1 | 4.2 |
| | 100 | 96.5 | 2.8 | 3.9% | 98.4 | 2.6 | 3.6 | 96.1 | 2.5 | 3.3 | 97.9 | 2.8 | 3.5 | 98.1 | 2.3 | 3.1 | 96.6 | 2.6 | 3.5 |

**Table 3.** *Cont.*

| OCP | Amount Added ng/mL | Tap Water | | | Ground Water | | | Municipal Water | | | Apple Juice | | | Litchi Juice | | | Pomegranate Juice | | |
|---|---|---|---|---|---|---|---|---|---|---|---|---|---|---|---|---|---|---|---|
| | | Extraction Yield | Intraday RSD (%) | Interday RSD (%) | Extraction Yield | Intraday RSD (%) | Interday RSD (%) | Extraction Yield | Intraday RSD (%) | Interday RSD (%) | Extraction Yield | Intraday RSD (%) | Interday RSD (%) | Extraction Yield | Intraday RSD (%) | Interday RSD (%) | Extraction Yield | Intraday RSD (%) | Interday RSD (%) |
| Endrin | 0.1 | 96.4 | 3.9 | 5.2 | 97.3 | 4.7 | 5.6 | 95.5 | 4.5 | 5.6 | 95.4 | 4.7 | 5.4 | 96 | 5.1 | 5.8 | 95.6 | 4.6 | 5.6 |
| | 1 | 97.1 | 3.6 | 4.9 | 98 | 4.6 | 5.2 | 96.1 | 4.1 | 4.8 | 96.6 | 4.5 | 4.9 | 97.2 | 4.7 | 5.2 | 96.4 | 3.7 | 5 |
| | 10 | 97.8 | 3.3 | 4.5 | 98.5 | 4.7 | 4.9 | 97.9 | 3.9 | 4.7 | 97.2 | 3.9 | 4.7 | 97.6 | 4.2 | 4.9 | 97.4 | 3.2 | 4.5 |
| | 100 | 98.7 | 3 | 3.9 | 98.7 | 3.6 | 4.1 | 98.5 | 3.5 | 4.2 | 97.8 | 2.6 | 3.3 | 98.1 | 3.2 | 3.9 | 98.7 | 2.9 | 3.9 |
| β-Endosulfan | 0.1 | 94.1 | 4.5 | 5.6 | 94.3 | 4.6 | 5.8 | 94.4 | 4.7 | 5.4 | 95.5 | 4.3 | 5.4 | 95.5 | 4.4 | 5.3 | 94.2 | 4.2 | 5.1 |
| | 1 | 95.5 | 3.6 | 4.7 | 95.6 | 4.3 | 5.2 | 95.1 | 4 | 4.8 | 96.8 | 4 | 5.1 | 96.3 | 3.9 | 4.8 | 95.9 | 3.7 | 4.8 |
| | 10 | 96.2 | 3.2 | 4.3 | 96.7 | 3.6 | 4.7 | 96.8 | 2.8 | 3.8 | 97.3 | 3.7 | 4.6 | 97.4 | 3.4 | 4.6 | 96.9 | 3.3 | 4.4 |
| | 100 | 97.8 | 2.9 | 3.8 | 97.8 | 2.5 | 3.6 | 97.4 | 2.4 | 3.4 | 98.8 | 2.6 | 3.5 | 98.6 | 2.2 | 3.1 | 97.2 | 2.4 | 3.3 |
| p,p' Dichlorodiphenyldichloroethane (p,p' DDD) | 0.1 | 96.4 | 3.7 | 5.3 | 97.9 | 4.6 | 5.5 | 96.7 | 4.1 | 5.5 | 95.8 | 4.7 | 5.8 | 96.9 | 4.9 | 5.6 | 95.5 | 4.6 | 5.5 |
| | 1 | 96.1 | 3.7 | 4.9 | 98.5 | 4.7 | 4.9 | 97.4 | 4 | 5.1 | 96.3 | 4.2 | 5.1 | 97.1 | 4.2 | 5.1 | 95.8 | 4.1 | 4.9 |
| | 10 | 97.1 | 3.4 | 4.3 | 98.9 | 4.5 | 4.6 | 97.8 | 3.8 | 4.6 | 97.5 | 3.7 | 4.6 | 97.6 | 3.7 | 4.6 | 96.4 | 3.8 | 4.7 |
| | 100 | 98.2 | 3.2 | 3.9 | 99.1 | 3.7 | 4.1 | 98.1 | 2.5 | 3.3 | 97.8 | 2.7 | 3.6 | 98.7 | 3.2 | 4.3 | 97.3 | 3.1 | 4.2 |
| Endrin Aldehyde | 0.1 | 95 | 4.2 | 4.1 | 95.2 | 4.7 | 5.7 | 94.1 | 4.8 | 5.6 | 94.1 | 4.6 | 5.5 | 95.2 | 4.8 | 5.8 | 95.8 | 4.4 | 5.5 |
| | 1 | 95.8 | 3.8 | 4.8 | 96.5 | 4.4 | 5.5 | 95.4 | 4.1 | 5 | 95.5 | 4.2 | 5.1 | 95.3 | 3.8 | 4.7 | 96.1 | 4 | 5.2 |
| | 10 | 96.4 | 3.1 | 4.1 | 97.8 | 3.5 | 4.4 | 95.7 | 3 | 4 | 96.4 | 3.3 | 4.5 | 96.7 | 3.5 | 4.4 | 96.8 | 3.5 | 4.7 |
| | 100 | 96.9 | 2.8 | 3.5 | 98.9 | 2.2 | 3.1 | 96.2 | 2.7 | 3.5 | 97.3 | 2.3 | 3.2 | 98.9 | 2.3 | 3.2 | 97.1 | 2.3 | 3.2 |
| Endosulfan sulfate | 0.1 | 95.5 | 4.4 | 5.6 | 96.4 | 4.8 | 5.6 | 96.4 | 4.9 | 5.8 | 94.6 | 5.2 | 5.9 | 95.2 | 5.1 | 5.9 | 94.2 | 4.9 | 5.8 |
| | 1 | 96.6 | 3.9 | 4.9 | 97.3 | 4.4 | 5.2 | 97 | 4.6 | 5.2 | 95.4 | 4.5 | 5.4 | 96.7 | 4.8 | 5.6 | 95.6 | 4.4 | 4.9 |
| | 10 | 97.7 | 3.5 | 4.6 | 98.4 | 4.1 | 4.9 | 97.9 | 3.2 | 4.4 | 96.2 | 3.9 | 4.5 | 97.3 | 4.1 | 4.9 | 96.5 | 3.9 | 4.6 |
| | 100 | 98.1 | 3.2 | 3.8 | 99.2 | 3.5 | 4.6 | 98.8 | 2.7 | 3.3 | 97.5 | 3.2 | 4.3 | 98.2 | 3.4 | 4.2 | 97.7 | 3.3 | 4.3 |
| p,p' Dichlorodiphenyltrichloroethane (p,p' DDT) | 0.1 | 95.2 | 3.9 | 4.6 | 95.5 | 4.9 | 5.8 | 94.7 | 4.9 | 5.8 | 94.8 | 4.7 | 5.6 | 94.2 | 5 | 5.9 | 94.6 | 4.8 | 5.7 |
| | 1 | 96.7 | 3.4 | 4.2 | 96.1 | 4.6 | 5.7 | 95.2 | 4.2 | 5.2 | 95.4 | 4.1 | 5 | 95.5 | 3.9 | 5.2 | 95.8 | 4.3 | 5.3 |
| | 10 | 96.1 | 2.9 | 3.7 | 96.8 | 3.7 | 4.6 | 96.5 | 3.2 | 4.1 | 96.3 | 3.4 | 4.4 | 96.6 | 3.4 | 4.5 | 96.1 | 3.9 | 4.9 |
| | 100 | 97.5 | 2.6 | 3.5 | 97.7 | 2.3 | 3.4 | 97.5 | 2.8 | 3.6 | 96.9 | 2.6 | 3.5 | 97.6 | 2.4 | 3.1 | 97.7 | 3 | 4.2 |
| Endrin ketone | 0.1 | 96.4 | 3.5 | 5 | 96.7 | 4.8 | 5.6 | 95.1 | 4.2 | 5.6 | 94.4 | 4.6 | 5.7 | 95.6 | 5.3 | 5.9 | 95.1 | 4.8 | 5.6 |
| | 1 | 97.4 | 3.1 | 4.8 | 97.2 | 4.6 | 5.2 | 96.5 | 3.9 | 5.2 | 95.5 | 4.1 | 5.3 | 95.8 | 4.4 | 5.5 | 95.6 | 4.3 | 5.1 |
| | 10 | 98 | 2.9 | 3.5 | 98.2 | 4.1 | 4.9 | 97.8 | 2.9 | 4.1 | 96.9 | 3.8 | 4.8 | 97.1 | 3.9 | 4.7 | 96.6 | 3.6 | 4.6 |
| | 100 | 98.9 | 2.6 | 3.2 | 98.8 | 3.5 | 4.2 | 98.3 | 2.6 | 3.4 | 97.2 | 2.9 | 3.8 | 98.1 | 3.4 | 4.5 | 98.1 | 3 | 4.1 |
| Methoxychlor | 0.1 | 97 | 3.4 | 4.8 | 95.7 | 4.6 | 5.9 | 95.2 | 4.1 | 5.2 | 94.3 | 4.8 | 5.9 | 95.7 | 5.1 | 5.8 | 95.5 | 4.9 | 5.8 |
| | 1 | 97.8 | 3 | 4.1 | 96.2 | 4.5 | 5.6 | 96.6 | 3.9 | 5.1 | 95.4 | 4.2 | 5.3 | 96.1 | 4.2 | 5.1 | 95.5 | 4.4 | 5.4 |
| | 10 | 98.2 | 2.6 | 3.6 | 97.2 | 3.6 | 4.8 | 97.2 | 2.7 | 3.9 | 96.8 | 3.5 | 4.6 | 96.8 | 3.6 | 4.8 | 96.1 | 3.7 | 4.8 |
| | 100 | 98.8 | 2.4 | 3.3 | 98.8 | 2.9 | 4.1 | 98.7 | 2.3 | 3.5 | 97.3 | 2.8 | 3.7 | 97.2 | 2.5 | 3.6 | 97.6 | 3.1 | 4.1 |

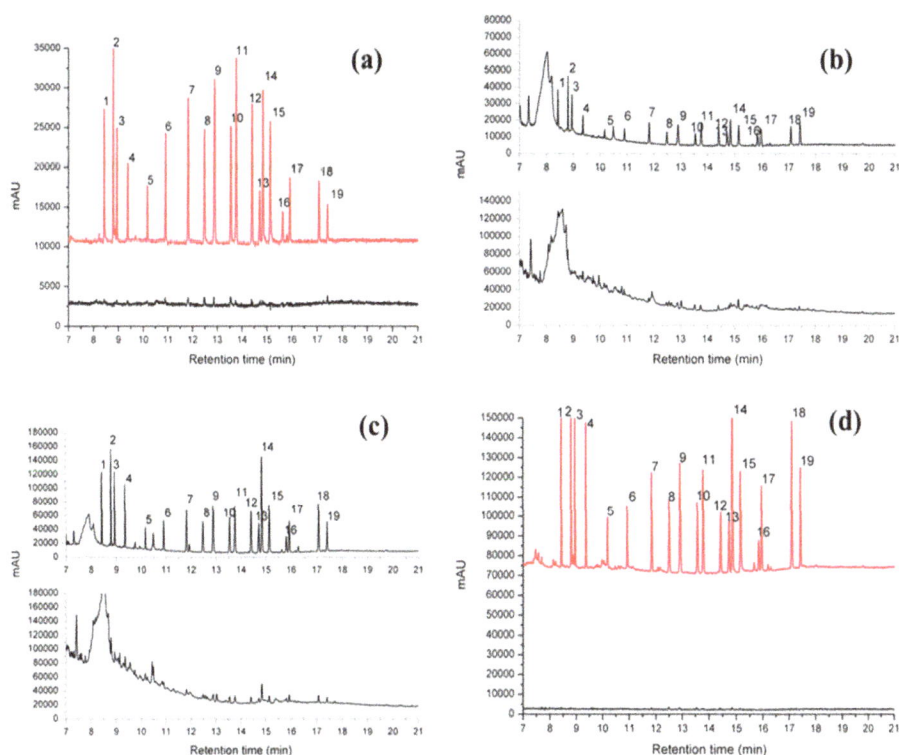

**Figure 3.** FPSE/GC-MS chromatograms of OCPs in real samples using sol-gel PEG-PPG-PEG coated FPSE media: (**a**) blank and spiked tap water; (**b**) blank and spiked apple juice; (**c**) blank and spiked pomegranate juice; (**d**) blank and spiked litchi juice. [Spiked concentration: 5 ppb].

## 2.3. Comparison with Other Reported Methods

The capability of the present method was compared with other extraction methods previously reported for the determination of OCPs, with the results being summarized in Table 3. The proposed method demonstrated satisfactory linearity, lower RSD values and comparable recoveries compared with the reported methods. The LOD values of present work were lower than those reported methods (Table 4). FPSE media have a large surface area that greatly increases the contact between analytes and sorbent. This in turn speeds up the extraction process with shorter extraction equilibrium time. Moreover, sol-gel FPSE media was directly inserted into the sample solution for extraction, and hence its usage was very simple. The sponge-like porous architecture of sol-gel sorbent and the capillary action of cellulose fabric synergistically diffuse the organic solvent into the sol-gel sorbent network during back-extraction and allows quantitative recovery of the extracted analytes even when a small volume of organic solvent is used. As such, FPSE also eliminates solvent evaporation and sample reconstitution, often considered to be an integral operation in solid phase extraction. A comparison between magnetic solid phase extraction and fabric phase sorptive extraction workflow is presented in Figure 4. As the schematic demonstrates, FPSE substantially simplifies the sample preparation workflow.

**Table 4.** Performance comparison between FPSE/GC-MS with other reported methods used in preconcentration and determination of organochlorine pesticides analytes.

| Sl. No. | Analyte | Matrix | Extraction Method | Chromatographic Technique | Linearity (ng/mL) | LOD (ng/mL) | RSD % | Reference |
|---------|---------|--------|-------------------|---------------------------|-------------------|-------------|-------|-----------|
| 1 | 8 OCPs | water | graphene SPE | GC-MS | 0.1–10 | 0.0019–0.0093 | <7.4 | [14] |
| 2 | 10 OCPs | Strawberry, strawberry jam, soil | SDME | GC-MS/MS | 0.5–50 | 0.002–0.150 | <15 | [16] |
| 3 | 9 OCPs | water | μ-SPE | GC-ECD | 0.1–100 | 0.0076–0.016 | <10 | [19] |
| 4 | 14 OCPs | water | HS-SBSE | GC-MS | 5–17 | 0.01–1.59 | <14.8 | [20] |
| 5 | 19 OCPs | water and juice samples | FPSE | GC-MS | 0.1–500 | 0.007–0.032 | <5 | Present work |

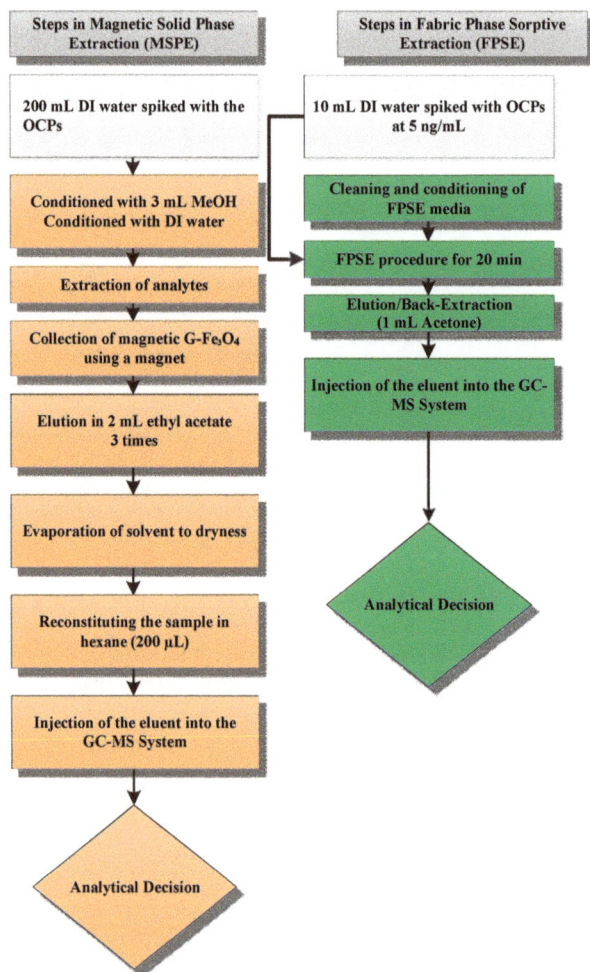

**Figure 4.** Comparison of sample preparation workflow between magnetic solid phase extraction (**left**) and fabric phase sorptive extraction (**right**).

## 3. Material and Methods

### 3.1. Reagents, Solvents and Material

Nineteen certified individual pesticide standards (purity, 96.8–99.55), including α-benzeneh exachloride, β-benzenehexachloride, γ-benzenehexachloride, δ-benzenehexachloride, heptachlor, aldrin, heptachlorepoxide, *trans*-chlordane, *cis*-chlordane, p,p dichlorodip henyldichloroethylene, dieldrin, endrin, β-endosulfan, p,p dichlorodip henyldichloroethane, endrin aldehyde, endosulfan sulphate, p,p dichlorodiphenyltrichloroethane, endrin ketone and methoxychlor were obtained from Sigma Aldrich (Bangalore, India) and were stored at −4 °C. Individual stock standard solutions (1 mg/L) of OCPs were prepared by dissolving an accurate weight of each pesticide in acetonitrile. Analytical grade methanol, hexane, acetone and acetonitrile were supplied by Merck (Mumbai, India). Water was deionized (Riviera, SCHOTT DURAN, Mainz, Germany) and filtered using 0.45-μm Nylon 6,6 membranes (Rankem, New Delhi, India) filtration assembly (Perfit, India). Working standard solutions were prepared daily by serial dilution of the individual stock solution with acetonitrile. An intermediate stock standard mixture was prepared by mixing the appropriate volumes of individual stock solutions and diluted with highly purified water to a required concentration.

### 3.2. Instrumentation

The pesticide analyses were performed using GC–MS QP 2010 plus (Shimadzu, Kyoto, Japan). Chromatographic separation was conducted with a fused silica capillary column Rtx-5MS, crossbonds 5% diphenyl and 95% dimethylpolysiloxane (30 m × 0.25 mm I.D., film thickness of 0.25 m, J & W Scientific, Folsom, CA, USA). Helium (purity ≥ 99.999%) was used as carrier gas at a constant flow of 1.0 mL/min. The temperature program was set initially at 100 °C for 2.5 min; ramp to 200 °C at a rate of 15 °C/min; 250 °C at 10 °C/min and finally to 300 °C at a rate of 6 °C/min being held for 2 min with total run time of 24 min. Injector temperature was maintained at 280 °C, and the injection volume was 1.0 μL in a splitless mode. Mass spectrometric parameters: electron impact ionization mode with an ionizing energy of 70 eV, injector temperature 250 °C, interface temperatures 230 °C, ion source temperature 200 °C. The mass spectrometer was operated in the selective ion monitoring (SIM) mode and the characteristic ions are given in Table 5. Full-scan data were acquired in the range of m/z 50–900 to obtain the fragmentation spectra of the analytes. The fragments of the ions monitored in SIM mode were selected based on good selectivity and high sensitivity.

**Table 5.** The performance characteristics of the proposed FPSE/GC-MS analytical method.

| Peak No. | OCP | Molecular Weight | CAS Number | Log Kow | Retention Time (min) | Qualitative Ion |
|---|---|---|---|---|---|---|
| 1 | α-Benzenehexachloride (α-BHC) | 290.83 | 319-84-6 | 3.81 | 8.43 | 183 *, 219, 109 |
| 2 | β-Benzenehexachloride (β-BHC) | 290.83 | 31-85-7 | 3.78 | 8.79 | 183 *, 219, 109 |
| 3 | γ-Benzenehexachloride (γ-BHC) | 290.83 | 58-89-9 | 3.72 | 8.94 | 183 *, 145, 109 |
| 4 | δ-Benzenehexachloride (δ-BHC) | 290.83 | 319-86-8 | 4.14 | 9.35 | 183 *, 219, 109 |
| 5 | Heptachlor | 373.32 | 76-44-8 | 6.10 | 10.17 | 100 *, 272, 237 |
| 6 | Aldrin | 364.90 | 309-00-2 | 6.50 | 10.92 | 66 *, 101, 263 |
| 7 | Heptachlorepoxide | 389.30 | 1024-57-3 | 5.40 | 11.83 | 53, 81 *, 353 |
| 8 | Trans Chlordane | 409.75 | 5103-74-2 | 6.16 | 12.49 | 176, 212 *, 375 |
| 9 | Cis Chlordane | 4.9.75 | 5103-71-9 | 6.16 | 12.90 | 237 *, 272, 373 |
| 10 | p,p Dichlorodiphenyldichlo roethylene (p,p′ DDE) | 318.02 | 72-55-9 | 6.51 | 13.56 | 176, 246 *, 318 |
| 11 | Dieldrin | 380.91 | 60-57-1 | 5.40 | 13.78 | 79 *, 263, 108 |
| 12 | Endrin | 380.90 | 72-80-8 | 5.20 | 14.42 | 81 *, 67, 263 |
| 13 | β-Endosulfan | 406.90 | 33213-65-9 | 3.62 | 14.72 | 207, 195 *, 159 |
| 14 | p,p Dichlorodiphenyldichlo roethane (p,p′ DDD) | 320.03 | 72-54-8 | 6.02 | 14.85 | 199, 235 *, 165 |
| 15 | Endrin Aldehyde | 380.89 | 7421-93-4 | 4.80 | 15.15 | 67 *, 250, 345 |
| 16 | Endosulfan sulfate | 422.90 | 1031-07-8 | 3.66 | 15.82 | 193, 207 *, 129 |
| 17 | p,p Dichlorodiphenyltrichlo roethane (p,p′ DDT) | 354.48 | 50-29-3 | 6.91 | 15.94 | 235 *, 165, 199 |
| 18 | Endrin ketone | 380.89 | 53494-70-5 | | 17.08 | 67 *, 281, 221 |
| 19 | Methoxychlor | 345.65 | 72-43-5 | 5.08 | 17.42 | 227 *, 169, 197 |

* most abundant ion.

### 3.3. Sample Collection and Preparation

Three types of environmental water samples, including tap water, ground water and municipal water, were collected and analyzed as part of real sample investigation. Tap water was taken from our lab faucet after flowing for 10 min. Ground water samples were collected from the bore well located within Punjabi University Campus, Patiala, Punjab, India in Pyrex borosilicate amber glass containers previously rinsed with triple-distilled water. No previous treatment was conducted for water samples, and all samples were stored at −4 °C in the refrigerator until analysis within 24 h.

The fruit juice (apple, litchi and pomegranate) samples were purchased from a local supermarket (Patiala, India). Fruit juice samples were stored at room temperature before use. Once opened, they were stored in specific food containers at 4 °C and analyzed within 2 days. A 20 mL aliquot of fresh juice was centrifuged at 4000 rpm for 15 min, and then the supernatant was filtered through a 0.45 μm membrane filter into a 50 mL conical flask. Before extraction, 10 mL of filtrate was diluted in a 1:1 ratio with deionized water in a 100 mL volumetric flask. After dilution, 10 mL of fruit juice was used for the extraction by FPSE procedure.

### 3.4. Preparation of Fabric Phase Sorptive Extraction Media

Taking the medium and low polarity of the organochlorine pesticides into consideration, two different sol-gel sorbent coatings, both on 100% cotton cellulose, were prepared and evaluated: sol-gel poly(ethylene glycol)-poly(propylene glycol)-poly(ethylene glycol) (sol-gel PEG-PPG-PEG) and sol-gel poly(caprolactone)-poly(dimethylsiloxane)-poly(caprolactone) (sol-gel PCAP-PDMS-PCAP). Both sol-gel PEG-PPG-PEG and sol-gel PCAP-PDMS-PCAP sorbents are moderately polar. Sol-gel sorbent coating on cellulose substrates involves a series of sequential steps, including (a) substrate selection and surface pre-treatment; (b) design and preparation of the sol solution for the sol-gel sorbent coating preparation of sol solution for coating; (c) sol-gel sorbent coating on the substrate via dip coating process; (d) condition, aging, and cleaning the sol–gel coated FPSE media; and (e) cutting the FPSE media into required size. A detailed procedure for the pre-treatment of cellulose substrate can be found in Kumar et al. [35]. The compositions and molar ratio between sol solution ingredients for sol-gel PEG-PPG-PEG and sol-gel PCAP-PDMS-PCAP, process of sol-gel coating, conditioning and aging as well as the post-coating cleaning protocols are given elsewhere [36]. Briefly, the cellulose fabric was treated first, with 1 M NaOH solution for an hour to eliminate all the residual finishing chemicals and to maximize the number of available hydroxide functional groups that would subsequently binds sol-gel sorbent network to the fabric surface via covalent bonding. The fabric was finally treated with 0.1 M HCl to neutralize any residual NaOH. The sol solution was prepared using the molar ratio sol-gel precursor: organic polymer: acetone: methylene chloride: trifluoroacetic acid: water at 1:0.02:3.26:3.74:1.25:3 for sol-gel PCAP-PDMS-PCAP and 1:0.13:1.94:2.3:0.75:3 for sol-gel PEG-PPG-PEG. The sol-gel sorbent coating on the fabric surface was created via dip coating process by immersing the fabric into the sol solution for 4 h. After sol-gel coating, the fabric was conditioned for 24 h at 50 °C. Subsequently, the coated fabric was cleaned with 50:50 (*v/v*) methanol and methylene chloride to remove any un-bonded sol solution ingredients. Finally, after drying, the sol-gel sorbent coated FPSE media were cut into 2.5 cm × 2.0 cm pieces.

### 3.5. Fabric Phase Extraction Procedure

The FPSE media (2.5 cm × 2.0 cm) was rinsed with acetone and then water before use to condition and equilibrate. It was then placed in a glass vial containing 10 mL pure water sample spiked with each of the OCPs at a concentration of 5 ng/mL. The magnetic stirrer was set at 900 rpm for 20 min, with the stirring being provided by a Teflon-coated magnetic stir bar inside the glass vial. After extraction, the FPSE media was removed from the sample solution and dried thoroughly with lint-free tissue. It was immediately placed in a glass vial containing 1 mL desorption solvent (acetone) for 15 min. The extract with target analytes was filtered with syringe filters prior to GC-MS analysis. One microliter of the extract was injected into the GC–MS system. After desorption, the extraction media was washed repeatedly with acetone and water for removing any possible residual analyte or other substances. To do this, the fabric phase media was transferred into 0.5 mL of acetone and then 0.5 mL water for 5 min to remove residual analytes. The carryover effect was randomly tested using 200 μL acetone followed by GC–MS. This examination clearly indicated that the device was reusable, as no analyte peaks were detected. A series of tests proved that the device was reusable up to 30 times without impacting on its extraction efficiency negatively.

## 4. Conclusions

In the present work, we reported the use of a cellulose fabric piece coated with sol-gel extraction sorbent as a sample preconcentration technique with the inherent features of both SPE and SPME methods. Here, the extraction phase, sol-gel PEG-PPG-PEG coated FPSE media with 100% cotton cellulose as the substrate demonstrated the highest affinity for the trace analysis of organochlorine pesticides in aqueous samples. This new sample preparation technique was further coupled to GC-MS for the determination of OCPs in different water and juice samples. Good analytical performance, including accuracy, precision and suitable detection limits with excellent linear dynamic ranges, was obtained under optimized conditions. It is evident from the results that the proposed FPSE/GC-MS methodology proved to be rapid, reliable, sensitive, time efficient, easy to implement using low sample and desorption solvent volume, providing excellent robustness and analytical reproducibility. The technology is expected to have promising potential for the routine analysis of organic pollutants, with the possibility of tuning the most selective sorbent coating based on the target compounds present at trace and ultra-trace level concentrations in various complex matrices.

**Author Contributions:** Data curation, R.K. (Ramandeep Kaur), R.K. (Ripneel Kaur), and S.R.; investigation, R.K. (Ramandeep Kaur), R.K. (Ripneel Kaur), and S.R.; methodology, A.K., K.G.F., A.K.M. and V.F.S.; project administration, A.K., K.G.F., A.K.M. and V.F.S.; supervision, A.K.M., K.G.F., A.K. and V.F.S.

**Funding:** This research was supported by the University Grant Commission (UGC), New Delhi through grant.

**Conflicts of Interest:** The authors declare no conflict of interest.

## References

1. Bajwa, A.; Ali, U.; Mahmood, A.; Chaudhry, M.J.I.; Syed, J.H.; Li, J.; Malik, R.N. Organochlorine pesticides (OCPs) in the Indus River catchment area, Pakistan: Status, soil-air exchange and black carbon mediated distribution. *Chemosphere* **2016**, *152*, 292–300. [CrossRef] [PubMed]
2. Oliveira, A.H.B.; Cavalcante, R.M.; Duaví, W.C.; Fernandes, G.M.; Nascimento, R.F.; Queiroz, M.E.L.R.; Mendonça, K.V. The legacy of organochlorine pesticide usage in a tropical semi-arid region (Jaguaribe River, Ceará, Brazil): Implications of the influence of sediment parameters on occurrence, distribution and fate. *Sci. Total Environ.* **2016**, *542*, 254–263. [CrossRef] [PubMed]
3. Sajid, M.; Basheer, C.; Narasimhan, K.; Buhmeida, A.; Al Qahtani, M.; Al-Ahwal, M.S. Persistent and endocrine disrupting organic pollutants: Advancements and challenges in analysis, health concerns and clinical correlates. *Nat. Environ. Pollut. Technol.* **2016**, *15*, 733–746.
4. Cai, S.; Sun, K.; Dong, S.; Wang, Y.M.; Wang, S.; Jia, L. Assessment of Organochlorine Pesticide Residues in Water, Sediment, and Fish of the Songhua River, China. *Environ. Forensics* **2014**, *15*, 352–357. [CrossRef]
5. Liu, W.X.; He, W.; Qin, N.; Kong, X.Z.; He, Q.S.; Ouyang, H.L.; Xu, F.L. The residues, distribution, and partition of organochlorine pesticides in the water, suspended solids, and sediments from a large Chinese lake (Lake Chaohu) during the high water level period. *Environ. Sci. Pollut. Res.* **2013**, *20*, 2033–2045. [CrossRef]
6. Rani, M.; Shanker, U.; Jassal, V. Recent strategies for removal and degradation of persistent & toxic organochlorine pesticides using nanoparticles: A review. *J. Environ. Manag.* **2017**, *190*, 208–222.
7. Tsygankov, V.Y.; Boyarova, M.D. Sample Preparation Method for the Determination of Organochlorine Pesticides in Aquatic Organisms by Gas Chromatography. *Achiev. Life Sci.* **2015**, *9*, 65–68. [CrossRef]

8.  Günter, A.; Balsaa, P.; Werres, F.; Schmidt, T.C. Influence of the drying step within disk-based solid-phase extraction both on the recovery and the limit of quantification of organochlorine pesticides in surface waters including suspended particulate matter. *J. Chromatogr. A* **2016**, *1450*, 1–8. [CrossRef]

9.  Hua, S.; Gong, J.L.; Zeng, G.M.; Yao, F.B.; Guo, M.; Ou, X.M. Remediation of organochlorine pesticides contaminated lake sediment using activated carbon and carbon nanotubes. *Chemosphere* **2017**, *177*, 65–76. [CrossRef]

10. Temoka, C.; Wang, J.; Bi, Y.; Deyerling, D.; Pfister, G.; Henkelmann, B.; Schramm, K.W. Concentrations and mass fluxes estimation of organochlorine pesticides in Three Gorges Reservoir with virtual organisms using in situ PRC-based sampling rate. *Chemosphere* **2016**, *144*, 1521–1529. [CrossRef]

11. Andrew, J.; Mahugija, M.; Henkelmann, B.; Schramm, K. Chemosphere Levels, compositions and distributions of organochlorine pesticide residues in soil 5–14 years after clean-up of former storage sites in Tanzania. *Chemosphere* **2014**, *117*, 330–337.

12. Rezaei, F.; Hosseini, M.R.M. New method based on combining ultrasonic assisted miniaturized matrix solid-phase dispersion and homogeneous liquid-liquid extraction for the determination of some organochlorinated pesticides in fish. *Anal. Chim. Acta* **2011**, *702*, 274–279. [CrossRef] [PubMed]

13. Bresin, B.; Piol, M.; Fabbro, D.; Mancini, M.A.; Casetta, B.; Del Bianco, C. Analysis of organo-chlorine pesticides residue in raw coffee with a modified "quick easy cheap effective rugged and safe" extraction/ clean up procedure for reducing the impact of caffeine on the gas chromatography-mass spectrometry measurement. *J. Chromatogr. A* **2015**, *1376*, 167–171. [CrossRef] [PubMed]

14. Lu, L.C.; Wang, C.I.; Sye, W.F. Applications of chitosan beads and porous crab shell powder for the removal of 17 organochlorine pesticides (OCPs) in water solution. *Carbohydr. Polym.* **2011**, *83*, 1984–1989. [CrossRef]

15. Shattar, S.F.A.; Zakaria, N.A.; Foo, K.Y. Feasibility of montmorillonite-assisted adsorption process for the effective treatment of organo-pesticides. *Desalin. Water Treat.* **2016**, *57*, 13645–13677. [CrossRef]

16. Han, Q.; Wang, Z.; Xia, J.; Xia, L.; Chen, S.; Zhang, X.; Ding, M. Graphene as an efficient sorbent for the SPE of organochlorine pesticides in water samples coupled with GC-MS. *J. Sep. Sci.* **2013**, *36*, 3586–3591. [CrossRef]

17. Huang, S.; He, S.; Xu, H.; Wu, P.; Jiang, R.; Zhu, F.; Ouyang, G. Monitoring of persistent organic pollutants in seawater of the Pearl River Estuary with rapid on-site active SPME sampling technique. *Environ. Pollut.* **2015**, *200*, 149–158. [CrossRef]

18. Fernandes, V.C.; Subramanian, V.; Mateus, N.; Domingues, V.F.; Delerue-Matos, C. The development and optimization of a modified single-drop microextraction method for organochlorine pesticides determination by gas chromatography-tandem mass spectrometry. *Microchim. Acta* **2012**, *178*, 195–202. [CrossRef]

19. He, Z.; Wang, P.; Liu, D.; Zhou, Z. Hydrophilic-lipophilic balanced magnetic nanoparticles: Preparation and application in magnetic solid-phase extraction of organochlorine pesticides and triazine herbicides in environmental water samples. *Talanta* **2014**, *127*, 1–8. [CrossRef]

20. Huang, Z.; Lee, H.K. Micro-solid-phase extraction of organochlorine pesticides using porous metal-organic framework MIL-101 as sorbent. *J. Chromatogr. A* **2015**, *1401*, 9–16. [CrossRef]

21. Zhou, Q.; Huang, Y.; Xiao, J.; Xie, G. Micro-solid phase equilibrium extraction with highly ordered $TiO_2$ nanotube arrays: A new approach for the enrichment and measurement of organochlorine pesticides at trace level in environmental water samples. *Anal. Bioanal. Chem.* **2011**, *400*, 205–212. [CrossRef] [PubMed]

22. Grossi, P.; Olivares, I.R.B.; de Freitas, D.R.; Lancas, F.M. A novel HS-SBSE system coupled with gas chromatography and mass spectrometry for the analysis of organochlorine pesticides in water samples. *J. Sep. Sci.* **2008**, *31*, 3630–3637. [CrossRef] [PubMed]

23. Rodil, R.; Popp, P. Development of pressurized subcritical water extraction combined with stir bar sorptive extraction for the analysis of organochlorine pesticides and chlorobenzenes in soils. *J. Chromatogr. A* **2006**, *1124*, 82–90. [CrossRef] [PubMed]

24. Kabir, A.; Furton, K.G.; Malik, A. Innovations in sol-gel microextraction phases for solvent-free sample preparation in analytical chemistry. *Trac—Trends Anal. Chem.* **2013**, *45*, 197–218.

25. Aznar, M.; Úbeda, S.; Nerin, C.; Kabir, A.; Furton, K.G. Fabric phase sorptive extraction as a reliable tool for rapid screening and detection of freshness markers in oranges. *J. Chromatogr. A* **2017**, *1500*, 32–42. [CrossRef] [PubMed]

26. Alcudia-León, M.C.; Lucena, R.; Cárdenas, S.; Valcárcel, M.; Kabir, A.; Furton, K.G. Integrated sampling and analysis unit for the determination of sexual pheromones in environmental air using fabric phase sorptive extraction and headspace-gas chromatography–mass spectrometry. *J. Chromatogr. A* **2017**, *1488*, 17–25. [CrossRef] [PubMed]

27. García-Guerra, R.B.; Montesdeoca-Esponda, S.; Sosa-Ferrera, Z.; Kabir, A.; Furton, K.G.; Santana-Rodríguez, J.J. Rapid monitoring of residual UV-stabilizers in seawater samples from beaches using fabric phase sorptive extraction and UHPLC-MS/MS. *Chemosphere* **2016**, *164*, 201–207. [CrossRef] [PubMed]

28. Huang, G.; Dong, S.; Zhang, M.; Zhang, H.; Huang, T. Fabric phase sorptive extraction: Two practical sample pretreatment techniques for brominated flame retardants in water. *Water Res.* **2016**, *101*, 547–554. [CrossRef]

29. Kabir, A.; Furton, K.G.; Tinari, N.; Grossi, L.; Innosa, D.; Macerola, D.; Locatelli, M. Fabric phase sorptive extraction-high performance liquid chromatography-photo diode array detection method for simultaneous monitoring of three inflammatory bowel disease treatment drugs in whole blood, plasma and urine. *J. Chromatogr. B* **2018**, *1084*, 53–63. [CrossRef] [PubMed]

30. Locatelli, M.; Kabir, A.; Innosa, D.; Lopatriello, T.; Furton, K.G. A fabric phase sorptive extraction-High performance liquid chromatography-Photo diode array detection method for the determination of twelve azole antimicrobial drug residues in human plasma and urine. *J. Chromatogr. B* **2017**, *1040*, 192–198. [CrossRef]

31. Lakade, S.S.; Borrull, F.; Furton, K.G.; Kabir, A.; Marcé, R.M.; Fontanals, N. Dynamic fabric phase sorptive extraction for a group of pharmaceuticals and personal care products from environmental waters. *J. Chromatogr. A* **2016**, *1456*, 19–26. [CrossRef]

32. Montesdeoca-Esponda, S.; Sosa-Ferrera, Z.; Kabir, A.; Furton, K.G.; Santana-Rodríguez, J.J. Fabric phase sorptive extraction followed by UHPLC-MS/MS for the analysis of benzotriazole UV stabilizers in sewage samples. *Anal. Bioanal. Chem.* **2015**, *407*, 8137–8150. [CrossRef]

33. Guedes-Alonso, R.; Ciofi, L.; Sosa-Ferrera, Z.; Santana-Rodríguez, J.J.; Del Bubba, M.; Kabir, A.; Furton, K.G. Determination of androgens and progestogens in environmental and biological samples using fabric phase sorptive extraction coupled to ultra-high performance liquid chromatography tandem mass spectrometry. *J. Chromatogr. A* **2016**, *1437*, 116–126. [CrossRef] [PubMed]

34. Samanidou, V.; Filippou, O.; Marinou, E.; Kabir, A.; Furton, K.G. Sol–gel-graphene-based fabric-phase sorptive extraction for cow and human breast milk sample cleanup for screening bisphenol A and residual dental restorative material before analysis by HPLC with diode array detection. *J. Sep. Sci.* **2017**, *40*, 2612–2619. [CrossRef] [PubMed]

35. Yang, M.; Gu, Y.; Wu, X.; Xi, X.; Yang, X.; Zhou, W.; Li, J. Rapid analysis of fungicides in tea infusions using ionic liquid immobilized fabric phase sorptive extraction with the assistance of surfactant fungicides analysis using IL-FPSE assisted with surfactant. *Food Chem.* **2018**, *239*, 797–805. [CrossRef] [PubMed]

36. Santana-Viera, S.; Guedes-Alonso, R.; Sosa-Ferrera, Z.; Santana-Rodríguez, J.J.; Kabir, A.; Furton, K.G. Optimization and application of fabric phase sorptive extraction coupled to ultra-high performance liquid chromatography tandem mass spectrometry for the determination of cytostatic drug residues in environmental waters. *J. Chromatogr. A* **2017**, *1529*, 39–49. [CrossRef]

**Sample Availability:** Samples of the compounds are available from the authors upon request.

*Article*

# Synthetic Peptide Purification via Solid-Phase Extraction with Gradient Elution: A Simple, Economical, Fast, and Efficient Methodology

Diego Sebastián Insuasty Cepeda [1], Héctor Manuel Pineda Castañeda [1],
Andrea Verónica Rodríguez Mayor [1], Javier Eduardo García Castañeda [2],
Mauricio Maldonado Villamil [1], Ricardo Fierro Medina [1] and Zuly Jenny Rivera Monroy [1,*]

[1]    Chemistry Department, Universidad Nacional de Colombia, Bogotá, Carrera 45 No 26-85, Building 451,
        office 409, Bogotá 11321, Colombia; dsinsuastyc@unal.edu.co (D.S.I.C.); hmpinedac@unal.edu.co (H.M.P.C.);
        anvrodriguezma@unal.edu.co (A.V.R.M.); mmaldonadov@unal.edu.co (M.M.V.);
        rfierrom@unal.edu.co (R.F.M.)
[2]    Pharmacy Department, Universidad Nacional de Colombia, Bogotá Carrera 45 No 26-85, Building 450,
        Bogotá 11321, Colombia; jaegarciaca@unal.edu.co
*      Correspondence: zjriveram@unal.edu.co; Tel.: +57-1-3165000 (ext. 14436)

Academic Editor: Victoria Samanidou
Received: 28 February 2019; Accepted: 26 March 2019; Published: 28 March 2019

**Abstract:** A methodology was implemented for purifying peptides in one chromatographic run via solid-phase extraction (SPE), reverse phase mode (RP), and gradient elution, obtaining high-purity products with good yields. Crude peptides were analyzed by reverse phase high performance liquid chromatography and a new mathematical model based on its retention time was developed in order to predict the percentage of organic modifier in which the peptide will elute in RP-SPE. This information was used for designing the elution program of each molecule. It was possible to purify peptides with different physicochemical properties, showing that this method is versatile and requires low solvent consumption, making it the least polluting one. Reverse phase-SPE can easily be routinely implemented. It is an alternative to enrich and purified synthetic or natural molecules.

**Keywords:** peptide; solid phase extraction (SPE); preparative purification; gradient elution; solid phase peptide synthesis

---

## 1. Introduction

Peptides and proteins are molecules with various biological activities and wide structural diversity [1]. Presently peptides are used for industrial applications such as drug manufacturing [2], cosmetics [3], food [4], and agricultural products [5], among others. Peptides can be obtained mainly via molecular biology [6], solution-phase chemical synthesis [7], solid-phase chemical synthesis [8], and from natural sources [9]. These methodologies involve different techniques for the purification or enrichment of intermediates or final products. When the purification of large quantities of peptide is required, the available methods have limitations, and the purification process has to be repeated several times. The final purity of the product depends on the method used and the nature of the sample (analyte and matrix). The main methods utilized for peptide purification are RP-HPLC chromatography, flash chromatography, ion-exchange chromatography, hydrophobic interaction chromatography, gel filtration chromatography, size exclusion chromatography, and hydrophilic interaction chromatography [10–14]. These methods allow obtaining high-purity products. However, they use up large amounts of solvent, and in some cases, produce low yields and require long purification times, which cause an increase in production costs.

Solid-phase extraction (SPE) is a technique used mostly for sample pretreatment and enrichment [14]. It has applications in the removal of impurities and isolation of analytes from complex biomatrices, such as blood and urine [15]. SPE methodologies were developed for waste treatment and environmental monitoring. The development of new methodologies based on SPE made this technique more versatile, allowing pretreatment of any kind of sample in a wide concentration range [15]. SPE chromatographic separation is based on the same principles as liquid chromatography (LC). Frontal chromatography is the main process in the extraction step, while displacement chromatography is the process that governs the analyte desorption [16]. The main criterion for selecting the chromatography mode is the analyte's physicochemical properties [15,16]. SPE is regarded as a separation method with advantages over other methods, allowing a variety of applications along with speed, reproducibility, and efficiency [15]. It is a versatile separation technique, and has become important in the last decades. The development of new stationary phases for different applications has grown as a research field [17–19].

Reverse-phase SPE (RP-SPE) is the technique most used. A sample dissolved in a polar mobile phase is loaded onto the column, and then the non-retained impurities are eluted by washing with the same polar mobile phase. The analyte is then eluted with a less polar mobile phase containing an organic modifier. This elution may be isocratic or gradient [14,16,18,19]. SPE method development depends on the analyte's physicochemical properties and concentration, the matrix, the stationary phase, and the detection system [16–18]. Herraiz et al. studied the separation of a mixture of synthetic peptides using SPE with different stationary phases. Retention follows the order CN < C2 < Phenyl < Cyclohexyl < C8 < C18, with more than 90% recovery [18]. Kulczykowska et al. purified fish plasma nanopeptides via SPE [20], and Kamysz [21] purified peptides, estatherin SV2, temporin and calcitermin A, using the RP-SPE methodology, obtaining products with high purity (95–97%) [22,23].

In the present investigation, a methodology for semi-preparative peptide purification using RP-SPE and gradient elution was developed. The quantity of organic modifier solvent (%B$_e$) for eluting the target molecule was calculated using the chromatographic profile of each crude product; specifically, its retention time was calculated taking into account the HPLC system dwell time and the column dead time. Thus, the obtained %B$_e$ allowed us to design the elution program. Using this method, synthetic peptides with different physicochemical properties such as length, hydrophobicity, and amino acid composition were purified, obtaining products with high purity. Our results showed that it is possible to purify significant amounts of peptide in one step with good yields and low solvent consumption, without specialized equipment.

## 2. Results and Discussion

Peptides derived from different proteins were synthesized by means of manual Solid Phase Peptide Synthesis (SPPS) using the Fmoc/tBu strategy [24,25]. Both the Fmoc group removal and the coupling reaction were monitored by means of the Kaiser test. The coupling reactions were carried out using a 5-molar excess of reagents with respect to the resin equivalents, and in some cases, it was necessary to repeat these reactions until the Kaiser test was negative. Some reactions could be incomplete because of steric hindrance and chain aggregation, generating undesired species that may hinder the purification of the synthesized product. In SPPS, it is possible to obtain crude peptides with several species, which influences the yield and purity of the final peptide.

In this investigation, a methodology for the purification of synthetic peptides via RP-SPE chromatography with gradient elution was developed. First, both (i) the HPLC system dwell time and (ii) the column dead time (at flow rate of 2.0 mL/min) were determined. Second, the crude peptide was analyzed by means of RP-HPLC, using a monolithic C18 column (50 × 4.6 mm) and an elution gradient of 5/5/50/100/100/5/5% solvent B (TFA 0.05% in ACN) in 0/1/9/9.5/11/11.5/15 min. Third, the purification method was designed from the crude peptide chromatographic profile. Thus, the peak corresponding to the target peptide was identified and the retention time (t$_R$) was determined. This value was corrected (t'$_R$) by subtracting the initial delay time (t$_i$ = 1.00 min), the column dead time ($t_o = 0.36$ min), and the HPLC dwell time ($t_D = 0.90$ min, measured as shown in Figure 1). Fourth, using the $t'_R$ and the gradient slope, the organic modifier concentration in the mobile phase needed to elute the peptide (%$B_e$) was calculated using Equation (1).

$$\%B_e = t'_R \times \left(\frac{\Delta B}{t_G}\right) + \%B_i \tag{1}$$

$$t'_R = t_R - (t_i + t_o + t_D) \tag{2}$$

$$\Delta B = (\%B_f - \%B_i) \tag{3}$$

where %$B_e$ corresponds to the percentage of the organic solvent in which the elution of the peptide is expected in the SPE, $t_R$ is the retention time of the chromatographic profile of the crude peptide, and $\Delta B$ and $t_G$ are the change of the %B and the gradient time, respectively.

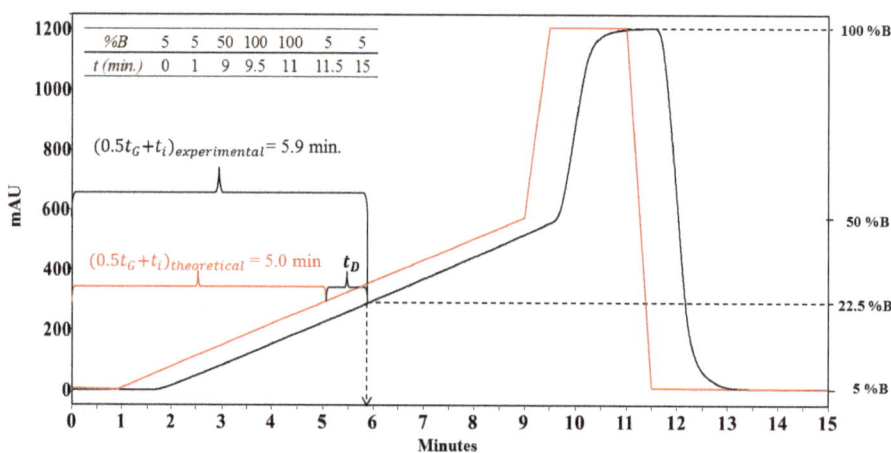

**Figure 1.** Dwell time determination [26–28]. Programed elution gradient (red line), experimental gradient performed by the HPLC system (black line). Delay time ($t_i$), gradient time ($t_G$), dwell time ($t_D$).

## 2.1. Purification of Synthetic Peptides via SPE

As an example, the purification process of the synthetic peptide: [20–25]LfcinB/[32–35]BFII: RRWQW RRLLR is shown, this sequence corresponds to a chimeric peptide derived from Lactoferricin B (LfcinB 20–25) and Buforin II (BFII 32–35). The crude peptide chromatographic profile (Figure 2) exhibits a main peak at $t_R = 5.10$ min, with a chromatographic purity of 60%. MS analysis showed that this peak had the expected molecular weight (data not shown). %$B_e$, in which peptide eluted, was calculated using Equation (1), as follows:

$$\%B_e = 2.84 \text{ min} \times \left(\frac{45\%B}{8 \text{ min}}\right) + 5\%B = 21\% \tag{4}$$

Then, 52.0 mg of crude peptide was dissolved in 1.0 mL of solvent A (0.05% TFA in water) and loaded onto a 5 g RP-SPE cartridge. The elution was performed by increasing the percentage of solvent B in the eluent. In Table 1, the design of the gradient elution program, taking into account the %$B_e$ obtained in Equation (4) (21% solvent B), is shown.

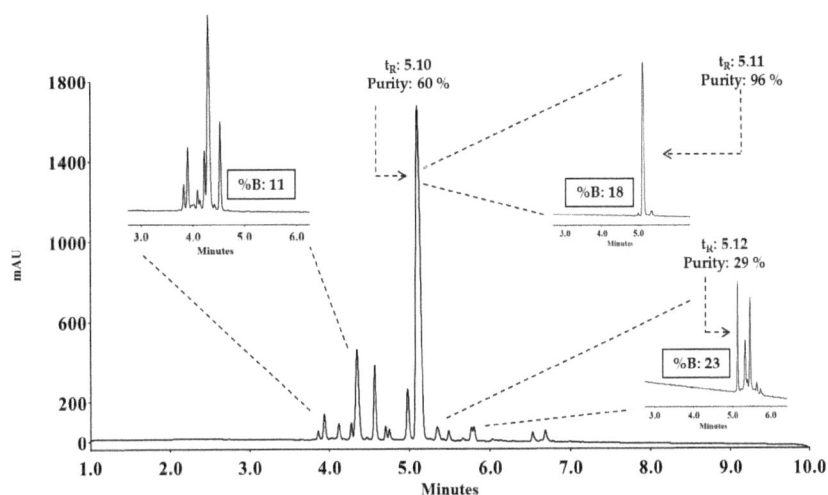

**Figure 2.** RP-HPLC analysis of crude $^{20-25}$LfcinB/$^{32-35}$BFII. The chromatographic profile shows a main peak at 5.10 min, corresponding to the peptide with a purity of 60%. The chromatographic profile of fractions N° 3, 6, and 11 collected during reversed-phase solid phase extraction (RP-SPE) purification is also shown. Specifically, those fractions contained 11, 18, and 23% solvent B, respectively.

**Table 1.** Program designed for the purification of $^{20-25}$LfcinB/$^{32-35}$BFII via RP-SPE. The framed fractions correspond to where the purest fraction probably elutes. The final volume of each fraction was 12 mL.

| Fraction N° | Solvent B | | Purity [b] (%) |
|---|---|---|---|
| | % | µL | |
| 1 | 0 | 0 | - |
| 2 | 5 | 600 | - |
| 3 | 11 | 1320 | - |
| 4 | 16 | 1920 | - |
| 5 | 17 | 2040 | 66 |
| 6 | 18 | 2160 | 96 |
| 7 | 19 | 2280 | 94 |
| 8 | 20 | 2400 | 92 |
| 9 [a] | 21 | 2520 | 88 |
| 10 | 22 | 2640 | 77 |
| 11 | 23 | 2760 | 23 |
| 12 | 24 | 2880 | - |
| 13 | 25 | 3000 | - |
| 14 | 50 | 6000 | - |
| 15 | 100 | 12000 | - |

[a] Fraction closest to calculated %B$_e$. [b] Chromatographic purity.

The analysis via RP-HPLC of collected fractions shows: (i) Fraction 3 contains the species corresponding to hydrophilic byproducts that are observed between 4–5 min in the crude profile (Figure 2. %B: 11). (ii) Fraction 6 (Figure 2. %B: 18) displayed a main peak with a $t_R$ of 5.11 min, which corresponds to the desired peptide with a chromatographic purity of 96%, revealing the great potential of this technique for the purification of synthetic peptides. Finally, (iii) fraction 11 has the hydrophobic byproducts that begin to elute and the peak that corresponds to the peptide with a purity of 29%. Fractions 6 to 8, which contain the peptide with purity greater than 90%, were mixed and lyophilized, obtaining 19.4 mg of purified product, the purification yield being 37%.

## 2.2. Purification of N-Glucosyl Amino Acids via SPE

For synthetizing N-glucopeptides using SPPS-Fmoc/tBu and building blocks methodology, it is necessary to obtain the intermediary Fmoc-Asn(GlcAc$_4$)-OtBu: Fmoc-L-Asn-(2,3,4,6-tetra-O-acetyl-β-D-N-glucopyranosyl)-OtBu. Briefly, the 2,3,4,6-tetra-O-acetyl-β-D-N-glucopyranosylamine was treated with Fmoc-Asp(OH)-OtBu (1:2 equivalents) that was previously activated with Dicyclohexylcarbodiimide (DIC) [29]. The final product was analyzed via RP-HPLC (Figure 3A) using an elution gradient of 5/5/100/100/5/5% solvent B in 0/1/18/20/20.5/24 min. The chromatographic profile presents two signals at 9.7 and 10.8 min that correspond to the Fmoc-Asp(OH)-OtBu (1) and the Fmoc-Asn(GlcAc$_4$)-OtBu (2), respectively (Figure 3A).

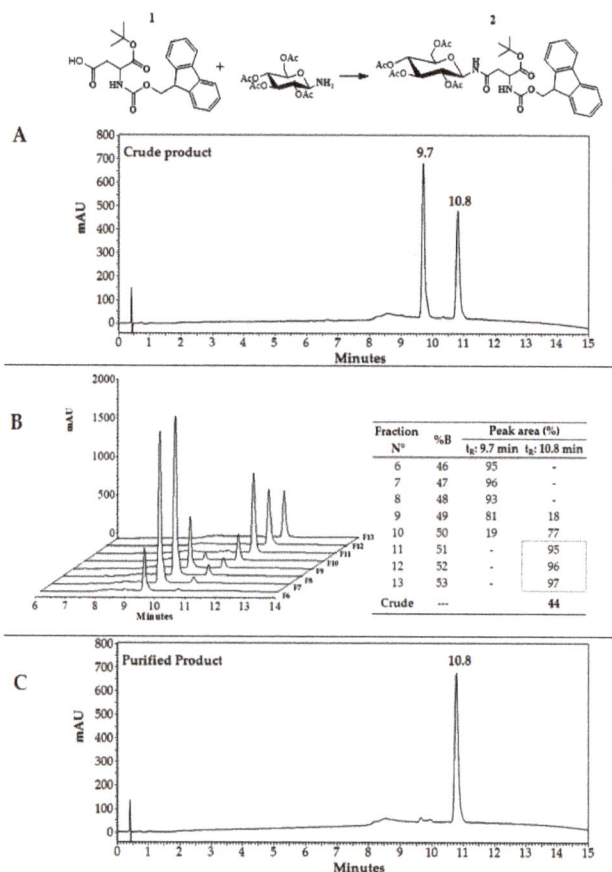

**Figure 3.** Purification of Fmoc-Asn(GlcAc$_4$)-OtBu (2) by RP-SPE. (**A**) Crude product chromatographic profile. (**B**) Chromatograms of collected fractions (6 to 13) (left) and used elution program (right). (**C**) Purified product chromatographic profile (fractions 11–13).

Analogously to peptide [20–25]LfcinB/[32–35]BFII, the purification of (2) was performed via RP-SPE as follows: Equation (1) was applied in order to predict the %B$_e$ required for elution of both species, it corresponds to 47% B for (1) and 53% B for (2). To illustrate, the purification process via RP-SPE, all fractions near the predicted %B were analyzed via RP-HPLC, and they are shown in Figure 3B.

In Figure 3B, it can be seen that the peak corresponding to (1) is present in fractions 6 to 8 with a percentage of area bigger than 90%. As the %B increases (F9–F10), the target molecule (2) begins to elute, progressively increasing its purity from 17% to 77%. From F11 to F13, the highest chromatographic purity for compound (**2**) was found (95–97%); those fractions are the closest to the predicted value of %B$_e$. Please observe that the peak at 9.7 min (**1**) was efficiently removed (Figure 3C). In this case, the purification yield of (**2**) was 44.5%.

We purified more than 100 synthetic peptides with different physicochemical properties using RP-SPE and gradient elution. This allowed us to routinely purify up to 150 mg peptide in a single step, obtaining high-purity products with excellent yields. Some examples of the peptides purified by our group are listed in Table 2.

Table 2. Peptides purified by RP-SPE.

| Peptide Code | Sequence | GRAVY [a] | %Ha [a] | Net Charge | $t_R$ | Purity [b] Crude | Purity [b] Purified | Purification Yield |
|---|---|---|---|---|---|---|---|---|
| 1 | Fc-*Ahx*-RLLR | N.D. | N.D. | +2 | 6.0 | 65 | 95 | 6 |
| 2 | Fc-*Ahx*-RLLRRLLR | N.D. | N.D. | +4 | 7.2 | 77 | 90 | 28 |
| 3 | *AcOx*-*Ahx*-RLLR | N.D. | N.D. | +2 | 4.4 | 46 | 99 | 10 |
| 4 | KKWQWK | −2.8 | 33 | +3 | 3.7 | 94 | 98 | 48 |
| 5 | IHSMNSTIL | 0.6 | 44 | +1 | 4.3 | 71 | 81 | 71 |
| 6 | PNNNKILVPK | −0.9 | 30 | +2 | 3.0 | 71 | 91 | 60 |
| 7 | LYIKGSGSTANLASSNYFPT | −0.1 | 30 | +1 | 4.9 | 55 | 72 | 16 |
| 8 | VSGLQYRVFR | −1.8 | 40 | +2 | 3.6 | 54 | 92 | 35 |
| 9 | N(Glc(Ac₄))-*Ahx*-RWQWRWQWR | N.D. | N.D. | N.D. | 6.0 | 64 | 76 | 61 |
| 10 | RWQWRWQWR-*Ahx*-N(Glc(Ac₄)) | N.D. | N.D. | N.D. | 6.4 | 60 | 66 | 70 |
| 11 | N(Glc(Ac₄))-*Ahx*-RWQWRWQWR-*Ahx*-N(Glc(Ac₄)) | N.D. | N.D. | N.D. | 6.6 | 66 | 80 | 13 |
| 12 | KKWQWKAKKLG | −1.8 | 36 | +5 | 3.9 | 89 | 99 | 63 |
| 13 | RRWQWRKKKLG | −2.5 | 27 | +6 | 3.8 | 91 | 99 | 66 |
| 14 | (RRWQWRKKKLG)₂-K-*Ahx* | −2.2 | 29 | +13 | 4.4 | 73 | 93 | 27 |
| 15 | Fc-*Ahx*-RRWQWR | N.D. | N.D. | +3 | 5.8 | 72 | 92 | 13 |
| 16 | AcFer-*Ahx*-RWQWRWQWR | N.D. | N.D. | +3 | 6.7 | 70 | 89 | 15 |
| 17 | RKKKMKKALQYIKLLKE | −1.2 | 35 | +7 | 4.9 | 61 | 86 | 7 |
| 18 | RYRRKKK | −3.8 | 0 | +6 | 0.7 | 93 | 99 | 41 |
| 19 | KMKKALQY | −1.1 | 37 | +3 | 3.1 | 84 | 98 | 42 |
| 20 | YIKLLKE | −0.1 | 42 | +1 | 4.2 | 99 | 99 | 26 |
| 21 | MKKALQYIKLLKE | −0.3 | 46 | +3 | 5.2 | 86 | 99 | 28 |
| 22 | FYFY | 0.8 | N.D. | 0 | 5.2 | 57 | 83 | 7 |
| 23 | KLLKKLLK | −0.1 | 50 | +4 | 4.0 | 90 | 99 | 55 |
| 24 | KLLK | −0.1 | N.D. | +2 | 1.6 | 89 | 92 | 53 |

%HA: Percentage of hydrophobic amino acids, GRAVY: Hydrophobicity. [a] Evaluated by prediction through the use of the online computer tool APD3: Antimicrobial Peptide Calculator and Predictor (http://aps.unmc.edu/AP/). [b] Chromatographic purity measured by area percentage. AcOx: Oxolinic acid, Fc: ferrocen motif, Glc(Ac₄): Tetra acetylated Glucose, Ahx: 6-aminohexanoic residue, AcFer: Ferulic acid.

RP-SPE is a supremely versatile technique because it makes possible to purify a great diversity of peptides with very varied physicochemical properties (Table 2). We purified hydrophilic molecules (sequences 18 and 24), hydrophobic ones (peptide 2), short and long chains (peptides 22 and 7), glycopeptides (peptides 9–11), peptides with chemical modifications (peptides 1–3, 15, and 16), and polyvalent peptides (sequence 14), among others. For the peptide purification process, we found yields ranging from 6% to 70% (Table 2). The efficiency and applicability of the purification process and the yield depends on the sample nature as: Solubility, quantity, composition of the crude product, presence of acid/basic species, the retention times of both, the impurities, and target peptide, among others. In Supplementary Material, it is shown the purification result of a challenging sample (Figure S1). Additionally, this technique allowed purifying small organic molecules, such as calix[4]resorcinarenes [30], as well as tetrahydrocannabinol from a natural source [31].

## 3. Materials and Methods

### 3.1. Reagents and Materials

The Rink amide resin and Fmoc-Gly-OH, Fmoc-Ala-OH, Fmoc-Val-OH, Fmoc-Leu-OH, Fmoc-Ile-OH, Fmoc-Phe-OH, Fmoc-Tyr(tBu)-OH, Fmoc-Trp(Boc)-OH, Fmoc-Ser(tBu)-OH, Fmoc-Thr(tBu)-OH, Fmoc-Cys(Trt)-OH, Fmoc-Met-OH, Fmoc-Asp(OtBu)-OH, Fmoc-Glu(OtBu)-OH, Fmoc-His(Trt)-OH, Fmoc-Lys(Boc)-OH, Fmoc-Arg(Pbf)-OH, Fmoc-Asn(Trt)-OH, Fmoc-Gln(Trt)-OH, Fmoc-Pro-OH, Fmoc-Ahx, N,N-Dicyclohexylcarbodiimide, and 1-Hydroxy-6-chlorobenzotriazole were purchased from AAPPTec (Louisville, KY, USA). The reagents such as, acetonitrile, trifluoroacetic acid, dichloromethane, diisopropylethylamine, N, N-dimethylformamide, ethanedithiol, isopropanol, methanol, and triisopropylsilane were purchased from Merck (Darmstadt, Germany). Supelclean™ SPE columns were purchased from Sigma-Aldrich (St. Louis, MO, USA), and Silicycle® SiliaPrep™ C18 columns were kindly donated by EcoChem Especialidades Químicas (Waterloo, QC, Canada).

### 3.2. Solid-Phase Peptide Synthesis (SPPS)

Peptides were synthesized using manual solid-phase peptide synthesis (SPPS-Fmoc/tBu) [24,25]. Briefly, Rink Amide resin (0.46 meq/g) was used as solid support. (i) Fmoc group removal was carried out through treatment with 25% 4-methylpiperidine in *N,N*-dimethylformamide (DMF). (ii) For the coupling reaction, Fmoc-amino acids (0.21 mmol) were pre-activated with DCC/1-Hydroxy-6-chlorobenzotriazole (6-Cl-HOBt) (0.20/0.21 mmol) in DMF at RT. (iii) Side chain deprotection reactions and peptide separation from the resin were carried out with a cleavage cocktail containing trifluoroacetic acid (TFA)/water/triisopropylsilane (TIPS)/ethanedithiol (EDT) (93/2/2.5/2.5% *v/v*). (iv) Crude peptides were precipitated by treatment with cool diethyl ether, dried at RT, and analyzed using RP-HPLC analytical chromatography.

### 3.3. Reverse-Phase High-Performance Liquid Chromatography (RP-HPLC) Analysis

RP-HPLC analysis was performed on a Chromolith® C-18 (50 × 4.6 mm) column using an Agilent 1200 liquid chromatograph (Omaha, NE, USA) with UV–Vis detector (210 nm). For the analysis of crude peptides (10 μL, 1 mg/mL), a linear gradient was applied from 5% to 50% solvent B (0.05% TFA in acetonitrile (ACN)) in solvent A (0.05% TFA in water) with a gradient time of 8 min. A delay time of 1.00 min was applied. The flow rate was 2.0 mL/min at RT. The column dead time was determined by injecting NaNO$_2$ (20 μL, 1 mg/mL) and isocratic elution with 5% B at a flow rate of 2.0 mL/min. Detection was at 210 nm. HPLC system dwell time was measured using the methodology described by Veronika Meyer with some modifications [28]. 0.8% Acetone in water was used as solvent B, and solvent A was 0.05% TFA in water. This analysis was performed without a column and using the following elution gradient: 5/5/50/100/100/5/5% B in 0/1/9/9.5/11/11.5/15 min. Flow rate was 2 mL/min, and detection was at 280 nm.

### 3.4. MALDI-TOF MS

The analysis was performed on an Ultraflex III MALDI–TOF mass spectrometer (Bruker Daltonics, Bremen, Germany) in reflectron mode, using an MTP384 polished steel target (BrukerDaltonics), 2,5-dihydroxybenzoic acid, or sinapinic acid as a matrix; Laser: 500 shots and 25–30% power.

### 3.5. Purification of molecules via RP-SPE

Peptides were purified using solid-phase extraction (SPE) on columns of two commercial houses (Silicycle® SiliaPrep™ C18 17%, 5 g, 45 μm, 60 Å and Supelclean™ SPE Tube 17%, 5 g, 45 μm, 60 Å). SPE columns were activated prior to use with 30 mL methanol, 30 mL ACN (containing 0.1% TFA), and were equilibrated with 30 mL water (containing 0.1% TFA). Up to 150 mg of crude peptide was dissolved in 1 to 2 mL of solvent A, and the solution was added to the column. The peptide elution was

*Molecules* **2019**, *24*, 1215

performed by increasing the percentage of solvent B in the eluent. Each fraction had a total volume of 12 mL. The collected fractions were analyzed via RP-HPLC and MALDI-TOF MS. The fractions containing the pure peptide were mixed and then lyophilized.

## 4. Conclusions

The RP-SPE methodology implemented is a very fast and efficient method for the purification of peptides. With this method, it is possible to achieve high purity and excellent yields at low cost. In this investigation, a new mathematical model based on the $t_R$ of the crude product was implemented in order to predict the percentage of organic modifier in which the peptide will elute.

More than 100 synthetic peptides with different physicochemical properties have been purified using this methodology. Furthermore, it was possible to purify organic molecules such as Fmoc-Asp(GlcAc$_4$)-OtBu, proving that this methodology is versatile and has advantages over other purification methods in terms of time and costs.

**Supplementary Materials:** The following are available online, Figure S1: Purification of a peptide containing seventeen amino acids.

**Author Contributions:** Z.J.R.M., M.M.V., R.F.M. and J.E.G.C. designed the peptides, purification protocols and supervised the research work. D.S.I.C. and H.M.P.C performed the peptide synthesis and characterization of those molecules by RP-HPLC and MALDI-TOF MS. A.V.R.M. synthesized and purified the Fmoc-Asp(GlcAc$_4$)-OtBu and the glucopeptides. All the authors contributed to anyzing the data and writing the manuscript.

**Funding:** This research was conducted with the financial support of División de Investigación y Extensión-sede Bogotá (DIEB), Universidad Nacional de Colombia (Project 41569: péptidos quiméricos derivados de la Lactoferricina bovina y la Buforina: diseño, síntesis y evaluación de su actividad antibacteriana).

**Acknowledgments:** Diego Insuasty and Héctor Pineda thanks to Dirección académica- Universidad Nacional de Colombia (Beca asistente docente), for financing his Ph.D. studies. Authors are thankful to Santiago Carrasquilla, EcoChem Especialidades Químicas, for donated the Silicycle® SiliaPrep™ C18 columns.

**Conflicts of Interest:** The authors declare no conflict of interest.

## References

1. Ratnaparkhi, M.P.; Pandya, C.S.P. Peptides and proteins in pharmaceuticals. *Int. J. Curr. Pharm. Res.* **2011**, *3*, 1–9.
2. Craik, D.J.; Fairlie, D.P.; Liras, S.; Price, D. The future of peptide-based drugs. *Chem. Biol. Drug Des.* **2013**, *81*, 136–147. [CrossRef] [PubMed]
3. Schagen, S.K. Topical peptide treatments with effective results. *Cosmetics* **2017**, *4*, 16. [CrossRef]
4. Sánchez, A.; Vázquez, A. Bioactive peptides: A review. *Food Qual. Saf.* **2017**, *1*, 29–46. [CrossRef]
5. Montesino, E.; Bardaji, E. Synthetic Antimicrobial peptides as agricultural pesticides for plant-disease control. *Chem. Biodivers.* **2008**, *5*, 1225–1237. [CrossRef]
6. D'Alessandro, R.; Masarone, D.; Buono, A.; Gravino, R.; Rea, A.; Salerno, G.; Golia, E.; Ammendola, E.; Del Giorno, G.; Santangelo, L.; et al. Natriuretic peptides: Molecular biology, pathophysiology and clinical implications for the cardiologist. *Future Cardiol.* **2013**, *9*, 519–534. [CrossRef]
7. Okada, Y.; Asama, H.; Wakamatsu, H.; Chiba, K.; Kamiya, H. Hydrophobic magnetic nanoparticle assisted one-pot liquid-phase peptide synthesis. *Eur. J. Org. Chem.* **2017**, *40*, 5961–5965. [CrossRef]
8. Kijewska, M.; Waliczek, M.; Cal, M.; Jaremko, L.; Jaremko, M.; Król, M.; Kołodziej, M.; Lisowski, M.; Stefanowicz, P.; Szewczuk, Z. Solid-phase synthesis of peptides containing aminoadipic semialdehyde moiety and their cyclisations. *Sci. Rep.* **2018**, *8*, 10462. [CrossRef]
9. Chandrudu, S.; Simerska, P.; Toth, I. Chemical methods for peptide and protein production. *Molecules* **2013**, *18*, 4373–4388. [CrossRef]
10. Aguilar, M.I. *Methods in Molecular Biology, HPLC of Peptides and Proteins, Methods and Protocols*; Humana Press Inc.: Totowa, NJ, USA, 2004.
11. Ren, H.L.J.; Zheng, X.Q.; Liu, X.L. Purification and characterization of antioxidant peptide from sunflower protein hydrolysate. *Food Technol. Biotechnol.* **2010**, *48*, 519–523.
12. Cytryniska, M.; Mak, P.; Zdybicka-barabas, A.; Suder, P.; Jakubowicz, T. Purification and characterization of eight peptides from Galleria mellonella immune hemolymph. *Peptides* **2007**, *28*, 533–546. [CrossRef]

13. Pingitore, E.V.; Salvucci, E.; Sesma, F.; Nader-Macias, M.E. Different Strategies for Purification of Antimicrobial Peptides from Lactic Acid Bacteria (LAB). In *Communicating Current Research and Educational Topics and Trends in Applied Microbiology*; Mendez Vilas, A., Ed.; FORMATEX: Badajoz, Spain, 2007; pp. 557–568.

14. Riding, G.A.; Wang, Y.; Walker, P.J. A Strategy for Purification and Peptide Sequence Analysis of Bovine Ephemeral Fever Virus Structural Proteins. In *Bovine Ephemeral Fever and Related Rhabdoviruses*; ACIAR Proceedings; St George, T.D., Uren, M.F., Young, P.L., Hoffman, D., Eds.; ACIAR: Canberra, Australia, 1992; Volume 44, pp. 98–102.

15. Poole, F. New trends in solid-phase extraction. *Trends Anal. Chem.* **2003**, *22*, 362–373. [CrossRef]

16. Hennion, M. Solid-phase extraction: Method development, sorbents, and coupling with liquid chromatography. *J. Chromatogr. A* **1999**, *856*, 3–54. [CrossRef]

17. Poole, F.; Gunatilleka, A.D.; Sethuraman, R. Contributions of theory to method development in solid-phase extraction. *J. Chromatogr. A* **2000**, *885*, 17–39. [CrossRef]

18. Herraiz, T.; Casal, V. Evaluation of solid-phase extraction procedures in peptide analysis. *J. Chromatogr. A* **1995**, *708*, 209–221. [CrossRef]

19. Palmblad, M.; Vogel, J.S. Quantitation of binding, recovery and desalting efficiency of peptides and proteins in solid phase extraction micropipette tips. *J. Chromatogr. B* **2005**, *814*, 309–313. [CrossRef] [PubMed]

20. Kulczykowska, E. Solid-phase extraction of arginine vasotocin and isotocin in fish samples and subsequent gradient reversed-phase high-performance liquid chromatographic separation. *J. Chromatogr. B Biomed. Sci. Appl.* **1995**, *673*, 289–293. [CrossRef]

21. Kamysz, W.; Okrój, M.; Lempicka, E.; Ossowski, T.; Lukasiak, J. Fast and efficient purification of synthetic peptides by solid-phase extraction. *Acta Chromatogr.* **2004**, *14*, 180–186.

22. Herraiz, T. Sample preparation and reversed phase-high performance chromatography analysis of food-derived peptides. *Anal. Chim. Acta* **1997**, *352*, 119–139. [CrossRef]

23. Causon, R.C.; Mcdowall, R.D. Sample pretreatment techniques for the bioanalysis of peptides. *J. Control. Release.* **1992**, *21*, 37–48. [CrossRef]

24. Vergel Galeano, C.F.; Rivera Monroy, Z.J.; Rosas Pérez, J.E.; García Castañeda, J.E. Efficient synthesis of peptides with 4-methylpiperidine as Fmoc removal reagent by solid phase synthesis. *J. Mex. Chem. Soc.* **2014**, *58*, 386–392.

25. Huertas, N.d.J.; Rivera Monroy, Z.J.; Medina, R.F.; García Castañeda, J.E. Antimicrobial activity of truncated and polyvalent peptides derived from the FKCRRQWQWRMKKGLA sequence against *Escherichia coli* ATCC 25922 and *Staphylococcus aureus* ATCC 25923. *Molecules* **2017**, *22*, 987. [CrossRef] [PubMed]

26. Kuss, H.J. Prediction of Gradients. In *The HPLC Expert: Possibilities and Limitations of Modern High Performance Liquid Chromatography*, 1st ed.; Kromidas, S., Ed.; Wiley-VCH: Weinheim, Germany, 2016; pp. 185–187.

27. Snyder, L.R.; Kirkland, J.J.; Dolan, J.W. *Introduction to the Modern Liquid Chromatography*, 3rd ed.; Wiley: Hoboken, NJ, USA, 2010.

28. Meyer, V.R. *Practical High-Performance Liquid Chromatography*, 5th ed.; John Wiley and Sons, Ltd.: Oxford, UK, 2010; pp. 373–385.

29. Pinzón-Martín, S.M.; Medina, R.F.; Iregui Castro, C.A.; Rivera Monroy, Z.J.; García Castañeda, J.E. Novel synthesis of N-glycosyl amino acids using T3P®: Propylphosphonic acid cyclic anhydride as coupling reagent. *Int. J. Pept. Res. Ther.* **2017**, *24*, 291–298. [CrossRef]

30. Castillo-Aguirre, A.; Rivera-Monroy, Z.; Maldonado, M. Selective o-alkylation of the crown conformer of tetra(4-hydroxyphenyl)calix[4]resorcinarene to the corresponding tetraalkyl ether. *Molecules* **2017**, *22*, 1660. [CrossRef]

31. Gallo-Molina, C.; Castro-Vargas, H.I.; Garzón-Méndez, W.F.; Martínez-Ramírez, J.A.; Rivera-Monroy, Z.J.; King, J.W.; Parada-Alfonso, F. Extraction, isolation and purification of tetrahydrocannabinol from the Cannabis sativa L. plant using supercritical fluid extraction and solid phase extraction. *J. Supercrit. Fluids* **2019**, *146*, 208–216. [CrossRef]

**Sample Availability:** Samples of the compounds used in this paper are available from the authors.

MDPI

St. Alban-Anlage 66

4052 Basel

Switzerland

Tel. +41 61 683 77 34

Fax +41 61 302 89 18

www.mdpi.com

*Molecules* Editorial Office

E-mail: molecules@mdpi.com

www.mdpi.com/journal/molecules

www.ingramcontent.com/pod-product-compliance
Lightning Source LLC
Chambersburg PA
CBHW051859210326

41597CB00033B/5956